X Mexican School on Gravitation and Mathematical Physics 2014

Reaching a Century: Classical and Modified General Relativity's Attempts to Explain the Evolution of the Universe

Journal of Physics: Conference Series Volume 654

Playa del Carmen, Mexico
1 – 5 December 2014

Editors:

Ricardo Becerril Barcenas
Hector Hugo Hernandez Hernandez
Miguel Sabido

ISBN: 978-1-5108-1728-9
ISSN: 1742-6588

Printed from e-media with permission by:

Curran Associates, Inc.
57 Morehouse Lane
Red Hook, NY 12571

CURRAN ASSOCIATES INC. proceedings.com

Some format issues inherent in the e-media version may also appear in this print version.

Printed by Curran Associates, Inc. (2016)

For permission requests, please contact the Institute of Physics at the address below.

Institute of Physics
Dirac House, Temple Back
Bristol BS1 6BE UK

Phone: 44 1 17 929 7481
Fax: 44 1 17 920 0979

techtracking@iop.org

Additional copies of this publication are available from:

Curran Associates, Inc.
57 Morehouse Lane
Red Hook, NY 12571 USA
Phone: 845-758-0400
Fax: 845-758-2634
Email: curran@proceedings.com
Web: www.proceedings.com

Table of contents

Volume 654

X Mexican School on Gravitation and Mathematical Physics: "Reaching a Century: Classical and Modified General Relativity's Attempts to explain de evolution of the Universe"
1–5 December 2014, Playa del Carmen, Quintana Roo, México

Accepted papers received: 15 October 2015
Published online: 11 November 2015

Preface

Papers

Preface

The collection of papers in this volume was presented during the X Mexican School on Gravitation and Mathematical Physics, which was held in Playa del Carmen, Quintana Roo, México, December 1-5, 2014. The Mexican School on Gravitation and Mathematical Physics is a series of conferences sponsored by the Mexican Physical Society that started in 1994 with the purposes of discussing and exchanging current ideas in gravitational physics. Each Mexican School has been devoted to a particular subject, and these have included supergravity, branes, black holes, the early Universe, observational cosmology, quantum gravity and numerical relativity.

In this ocasion the theme of the school was *Reaching a Century: Classical and Modified General Relativity's Attempts to explain the evolution of the Universe*, which focused on the discussion of classical and modified aspects of general relativity. Following our previous Schools, world leaders in the field were invited to give courses and plenary lectures. More specialized talks were also presented in parallel sessions, and some of them have been included in these proceedings.

The contributions in this volume have been reviewed and represent some of the courses, plenary talks and contributed talks presented during our X School. We are indebted to the contributors of these proceedings as well as to the rest of the participants in our Mexican School all for making of it a complete success.

As for financial support we should mention the Mexican National Science and Technology Council (CONACyT), the Royal Society of London (UK), the Mexican Physical Society (SMF), as well as several Institutions including: Centro de Investigación y Estudios Avanzados (CINVESTAV), Universidad Autónoma Metropolitana Iztapalapa (UAM-I), Universidad de Guanajuato (UG), and Universidad Nacional Autónoma de México (UNAM).

Ricardo Becerril Bárcenas, Héctor Hugo Hernández Hernández, Miguel Sabido
Editors

Peer review statement

All papers published in this volume of *Journal of Physics: Conference Series* have been peer reviewed through processes administered by the proceedings Editors. Reviews were conducted by expert referees to the professional and scientific standards expected of a proceedings journal published by IOP Publishing.

Time series analysis of Gravitational Wave signals using neural networks

M. Carrillo, J. A González, M. Gracia-Linares, F. S. Guzmán
Instituto de Física y Matemáticas, Universidad Michoacana de San Nicolás de Hidalgo. Edificio C-3, Cd. Universitaria, 58040 Morelia, Michoacán, México.

Abstract. We present an algorithm based on neural networks that predicts the mass ratio in a binary black hole collision out of given Gravitational Wave (GW) signals. In this brief analysis, the network is trained with a small sample of GW signals generated with numerical simulations. The effectiveness of the algorithm is evaluated with GW generated again with simulations with given mass ratios unknown to the network and found to be in the worst scenarios less than 10%.

1. Introduction

The sources of Gravitational Waves (GWs) are of garden variety, from sources such as binary systems involving black holes and neutron stars, supernovae core collapse, down to high energy accretion processes [1]. Compact binary systems are the most promising ones to be detected [2], because of the large expected event rate [3].

The set of parameters characterizing a binary black hole collision involves the parameters of the orbit before the collision, the orientation of the independent spins of the black holes, their mass ratio and an extraordinary amount of parameters involved if matter surrounding the system are taken into account. Speeding up the estimation of these parameters is a matter of the major importance because of the amount of computational time for the signal analysis and filtering. Parameter estimation counts with efficient techniques, including for instance the use of Markov chains [4] and interpolation of waveforms [5], which have shown to be very efficient. However, it is still important to reduce the computational costs of estimations.

The characterization of the GW source involves the analysis of the properties of the signal (assuming this has been filtered) involves inverse problem strategies. As a first step, and in order to investigate whether neural networks can help at the estimation of parameters, we introduce in this paper an algorithm based on neural networks that is capable to predict the parameters of the source. At this stage we still do not consider noise filtering or statistical inference and only assume GW signals are given, and the goal is to give an approximate of the parameters value. At this point we also are wondering about the precision of the parameter prediction and will worry later about the efficiency.

We restrict to the case of binary black hole collisions and only one parameter, namely the mass ratio between the black holes. We generate GW signals using the Einstein Toolkit code [6] to generate GW signals that we use to train our network. Then provided a new signal, also generated through simulations, we challenge the network to guess the black holes mass ratio. This allows us to estimate the effectiveness of the network and the possibilities it has to tackle problems involving more parameters.

The paper is organized as follows. In section 2 we present a general description of artificial neural networks, in section 3 we present the specific properties of the simulations used to train the network, in section 4 we show the efficiency of the algorithm and in section 5 we draw some conclusions and perspectives.

2. Neural network description

Artificial Neural Networks (ANN) arose as an attempt to emulate the nervous system in both organization and stimuli processing from internal and external environment of living organisms [7]. Basically, a neural network is fed with some variables or parameters of a particular problem, and after a series of calculations it spits out a concrete outcome correlated with the input data. This kind of artificial intelligence is commonly used in pattern classification and recognition, which makes it specially appropiate to analyze the problem we intended to engage here.

The first notions on artificial neural network were established in 1943 by Warren McCulloch and Walter Pitts [8], creating a mathematical model for simple computational logical operations. Neural networks were proposed, in analogy to their biological counterparts, as a set of interconnected processing units called neurons, responsible for all calculations and data transfer, as described in Figure 1. A neuron can be defined with four components:

(i) Input nodes or synapses. A neuron has a determined number of nodes, depending on input connections. The value from each node x_i is weighted by a determined value w_i, that will vary according to the relevance of the connection.

(ii) Transfer function. It handles the income information from synapses, mapping a vector to a real number, it usually is the sum over all the weighted inputs.

(iii) Activation function. It takes the outcome from transfer function to a determined range through a certain mathematical function. The most popular activation functions, see figure 2, are the sigmoid function:

$$y(x) = \frac{1}{1+e^{-x}} \tag{1}$$

$$\frac{\partial y}{\partial x} = y(1-y) \tag{2}$$

or logistic function and the hyperbolic tangent:

$$y(x) = \tanh x \tag{3}$$

$$\frac{\partial y}{\partial x} = 1 - y, \tag{4}$$

however one may propose other functions that suit the problem, like sinusoidal, heaviside or even a linear function [9].

(iv) Connections. Deliver the neuron's outcome in a certain layer to neurons in the next layer.

Neural networks may have a lot of branches, depending on the number of neurons, but essentially are catalogued into *recurrent* and *feedforward networks* or non-recurrent. In recurrent networks, loops between neurons are allowed; meanwhile in a non-recurrent configuration, information flows only in one direction. Here we are focused on feedforward networks, since they are more widespread than recurrent networks, and can be used in countless applications despite their relative simplicity. The most common feedforward network is the multilayer perceptron (MLP), derived from the perceptron developed by Rosenblatt in 1958 [10]. The MLP is a fully connected network formed by 3 layers integrated by an arrange of neurons:

- Input Layer. This layer gets the parameters or patterns from outside, and pass them out over the next layer. For historical reasons, neurons at this level do not make calculations at all, and are used just for data transmission.

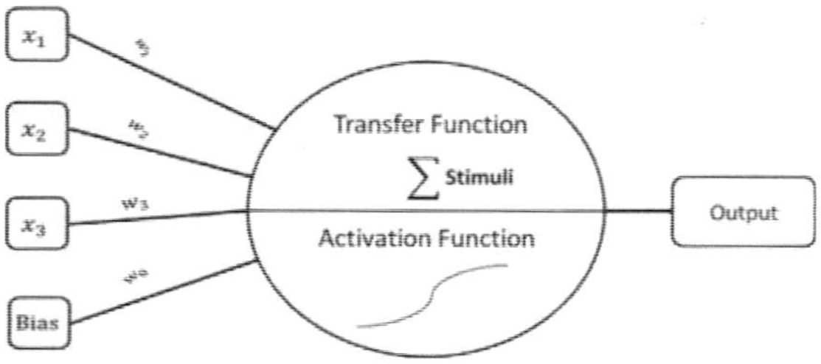

Figure 1. Scheme of an artificial neuron

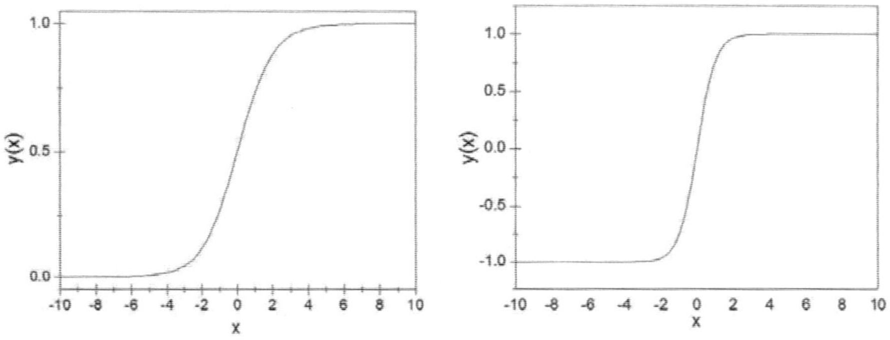

Figure 2. Most common activation functions in an artificial neuron: sigmoid (left) and hyperbolic tangent (right).

- Hidden Layer. Neurons receive the information from every neuron in the input or any other hidden layer if exists, even though the universal approximation theorem states that one layer is sufficient to approximate any function with a finite number of discontinuities, as long as the activation functions are non linear [11].
- Output Layer. Once data is received and operations are finished in this layer, an output pattern is shown and results can be analyzed.

2.1. Mathematical Representation of Neural Network

In order to illustrate a neural network in organization and operation in a mathematical fashion, let us suppose a 3 layer feedforward network with n inputs, m hidden neurons and l outputs. Since in the input layer no calculations are made, the j hidden neuron has the input vector $X_j = \{x_{1j}, \ldots, x_{ij}, \ldots, x_{nj}\}$ and its corresponding weight vector $W_{ij} = \{w_{1j}, \ldots, w_{ij}, \ldots, w_{nj}\}$, where i is referred to an input neuron. Then the transfer function maps the input vector from R^n to R, taking the linear combination of input vectors with the elements of the weight vector W_{ij} as coefficients:

$$\eta = \sum_{i=1}^{n} x_{ij} w_{ij} + w_0 \tag{5}$$

where w_0 is the weight of an extra synapse called *bias* with constant input equal to 1, added to shift the transfer function result or to avoid the particular case of a zero vector, in many cases it can be omitted for simplicity. Immediately after, the activation function F enters in action

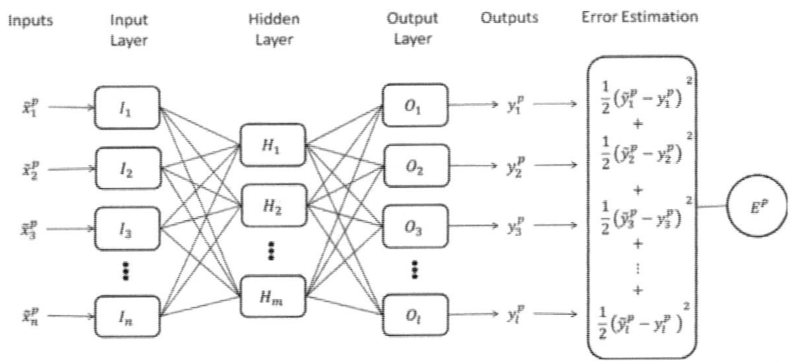

Figure 3. Scheme of the structure of a neural network and the usual estimation of error

giving the output σ_j:

$$\sigma_j = F(\eta) = F\left(\sum_{i=1}^{n} x_{ij}w_{ij} + w_0\right) \tag{6}$$

Finally, for the k output neuron we have the result:

$$y_k = G\left(\sum_{j=1}^{m} \sigma_j\theta_j + \theta_0\right) = G\left(\sum_{j=1}^{m}\left(F_j\left(\sum_{i=1}^{n} x_{ij}w_{ij} + w_0\right) * \theta_j\right) + \theta_0\right) \tag{7}$$

where G is the activation function and the θ_j are the weight coefficients corresponding to the input vector of the k output neuron.

2.2. *ANN Training*

In order to obtain the expected output for given particular pattern, one must get the appropriate configuration and weight matrix in the ANN. The way to do this is by changing all the synaptic weights through a "*training*". Suppose we have a set of N training patterns $\{\tilde{x}^p, \tilde{y}^p\}$, where p is the pattern number, $\tilde{x}^p$ and $\tilde{y}^p$ are the input and target output vectors respectively. In training phase, one seeks to reduce the error between the result y^p thrown by the ANN and the truly desired output $\tilde{y}^p$, figure

The most common way to measure this difference is by taking the sum of squared errors in output neurons:

$$E^p = \frac{1}{2}\sum_k \left(\tilde{y}_k^p - y_k^p\right)^2 \tag{8}$$

where k is the index of the output neurons. The overall performance or global error is defined by:

$$E^G = \sum_p E^p = \sum_p\sum_k \frac{1}{2}\left(\tilde{y}_k^p - y_k^p\right)^2 \tag{9}$$

2.2.1. Backpropagation Algorithm One of the most popular learning algorithms due to its acceptable performance in a great variety of problems, and its easily implementation in multilayer perceptron networks is the *backpropagation learning algorithm.* This method tries to search minimize the error function in terms of synaptic weights:

$$\nabla E = \left(\frac{\partial E}{\partial w_1}, \frac{\partial E}{\partial w_2}, \ldots, \frac{\partial E}{\partial w_k} \right) \rightarrow 0 \tag{10}$$

The purpose is to change every w_k for next time step $t+1$, propagating the error from the outputs to the input neurons at time t by the rule:

$$w_k(t+1) = w_k(t) + \triangle w_k(t) = w_k(t) - \gamma \frac{\partial E}{\partial w_k(t)} \tag{11}$$

where $\triangle w_k(t) = -\gamma \frac{\partial E}{\partial w_k(t)}$ and $0 < \gamma < 1$ the learning constant. Continuing with our previous example, we have for the connection from neuron i to j:

$$\Delta w_{ij} = -\gamma \sum_{k=1}^{l} \frac{\partial E_k}{\partial y_j} \frac{\partial y_j}{\partial s_j} \frac{\partial s_j}{\partial w_{ij}} = \gamma \delta_{ij} y_i \tag{12}$$

with $s = \sum_i^m w_{ij} y_i$, $\delta_{ij} = -\frac{\partial E}{\partial s_j}$ and $\frac{\partial s_j}{\partial w_{ij}} = y_i$. Considering F as the activation function, then $y = F(s)$ and δ_{ij} is calculated as:

$$\delta_{ij} = \begin{cases} (\tilde{y}_k - y_k)\frac{\partial F}{\partial s_j}\frac{\partial s_j}{\partial w_{ik}}, & \text{if } j = k, k \in \text{Output neurons.} \\ 0, & \text{if } j \neq k. \\ \sum_{k=1}^{l} \left[\delta_{jk} w_{jk} \right] \frac{\partial F}{\partial s_j}\frac{\partial s_j}{\partial w_{ij}}, & \text{if } j = h, h \in \text{Hidden neurons.} \end{cases} \tag{13}$$

In case a sigmoid is used as activation function, see eq. (2), then eq. (13) becomes:

$$\delta_{ij} = \begin{cases} (\tilde{y}_k - y_k) y_k (1 - y_k), & \text{if } j = k, k \in \text{Output neurons.} \\ 0, & \text{if } j \neq k. \\ \sum\limits_{k=1}^{l} \left[(\tilde{y}_k - y_k) y_k (1 - y_k) w_{jk} \right] y_j (1 - y_j), & \text{if } j = h, h \in \text{Hidden neurons.} \end{cases} \tag{14}$$

2.2.2. Scaled Conjugated Gradient Algorithm The disadvantages of backpropagation come from the choice of the learning rate, momentum, number of trainings and the possibility of getting stuck in a local minimum. That is why new algorithms in neural networks have emerged, like the scaled conjugated gradient [12], which is more efficient at approaching the minimum and without a dependency to a learning rate parameter. However, it is not excluded from the choice of hidden neurons or iterations. This method is based on the hipothesis that the minimum of the fuctions are locally almost quadratic. The weight change is determined by:

$$w_{t+1} = w_t + \alpha d_t \tag{15}$$
$$d_t = -g_t + \beta_t d_{t-1} \tag{16}$$

with g_t and β_t defined as:

$$g_t = \nabla E(w)|_{w=w_t} \tag{17}$$

$$\beta_t = \frac{g_t^T g_t}{g_{t-1}^T g_{t-1}} \tag{18}$$

where d_t y d_{t-1} are the conjugate directions in succesive iterations and T defines the transverse vector. The step value is set by the a_t coefficient and the minimum search direction by b_t. The a_t value is calculated as:

$$\alpha_t = -\frac{d_t^T g_t}{\delta_t} \quad (19)$$
$$\delta_t = d_t^T H_t d_t + \lambda_t ||d_t||^2 \quad (20)$$

λ_t is the scale coefficent. In case the error function is not quadratic, λ and δ can be negative and the weight increment can lead to an error increase. A sufficiently high value of λ guarantees that the Hessian matrix H_t is positive definite. In the other case, if $\delta < 0$, λ must be increased so that δ be greater than 0. A scale coefficient for $\bar{\delta}$ is suggested:

$$\bar{\lambda} = 2\left(\lambda - \frac{\delta_t}{||d_t||^2}\right) \quad (21)$$

And the new value for δ is:

$$\bar{\delta_t} = \delta_t + (\bar{\lambda_t} - \lambda_t)||d_t||^2 \quad (22)$$

This scaling also needs an adjustment to validate the quadratic approximation through:

$$\Delta_t = 2\frac{E(w_t) - E(w_t + \alpha_t d_t)}{\alpha_t d_t^T g_t} \quad (23)$$

If $\Delta_t > 0.75$ then the approximation is sufficiently good, and the value of λ_t can be reduced, proposing $\lambda_{t+1} = \lambda_t/2$. On the contrary, if $\Delta_t < 0.25$, $\lambda_t = 4\lambda_t$.

3. Parameter recognition in gravitational waves

The first problem on the implementation of a neural network, is how to enter data, namely defining the meaning the inputs in the neural network. In this case, the data used is the gravitational wave signal generated by binary black hole simulations. The number of inputs in the neural network is a set of sampled points, and the value entered in each input corresponds to the amplitude of the detected signal at time step t. This needs a great number of inputs to include at least the most significant part of the simulated signal, so that the GWs must be within an appropiate range in all cases.

As mentioned before, as a first step, we only consider a single parameter, the mass ratio between the black holes, while all other possible parameters are not considered. Unlike the real gravitational wave signals, with simulations it is possible to provide noise free signals with an arbitrary time resolution. The duration of the simulations is way to longer than the required information, and we only sample 300 time steps from the raw data, covering the region where the signature of the merger of the black holes on the gravitational waves are mostly present. In Figure 4 we show a sample of the GW signals used as input for the network.

These signals were generated with the Einstein Toolkit Cactus based code [6] considering the initial separation between black holes to be that of the first quasi-circular orbit [14] in order to obtain a quick merger for various mass ratios between the two black holes, ranging from 1 to 3.

The signal corresponds to the $l = 2$, $m = 0$ mode of the real part of the Ψ_4 Weyl scalar measured at $r = 30M$. As we can see, the segment where the GWs are appreciated, the signals are similar and the main difference comes up from the amplitude of the signal and a slight shift to the right in the same time segment. Then, the main function of the neural network will be to adjust itself to this differences and decrease the error through training with an appropiate number of GWs samples. The neural networks deployed were trained with 10 GWs samples with different mass ratios: 1.00, 1.10, 1.30, 1.40, 1.50, 1.60, 1.70, 2.10, 2.20, 2.30, and the following configurations:

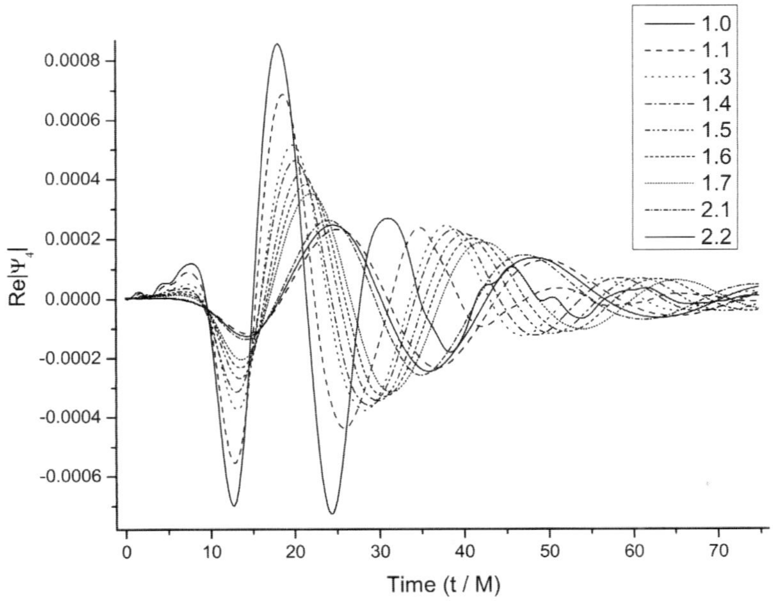

Figure 4. Some of the gravitational waves used for training the neural networks. See how similar they are in shape due to the similar mass ratio.

(i) 300 input neurons.

(ii) Run over a range from 200 to 800 hidden neurons with hyperbolic tangent as activation function with increments of 100 neurons. To measure the dependency of the number of hidden neurons, four neural networks with initial random weights have been used for each configuration.

(iii) An output neuron with a sigmoid activation function for the mass ratio output prediction.

After the training we provide GW signals generated using simulations for a given mass ratio unknown to the network, that has to be predicted by the network. Considering we know the mass ratio used to generate the test GWs it is possible to measure the error in the network's prediction.

4. Results

We thus estimate the mass ratio out of a given GW signal. In order to test the network we used 4 GWs corresponding to 1.20, 1.324, 1.9876, 2.2075 mass ratios shown in Figure 5. The algorithm implemented here is the scaled conjugate gradient, because it does not need to adjust a learning rate and momentum, leaving to explore just the number of trainings and hidding neurons. Given that the number of training samples is low, one can have hidden neurons in

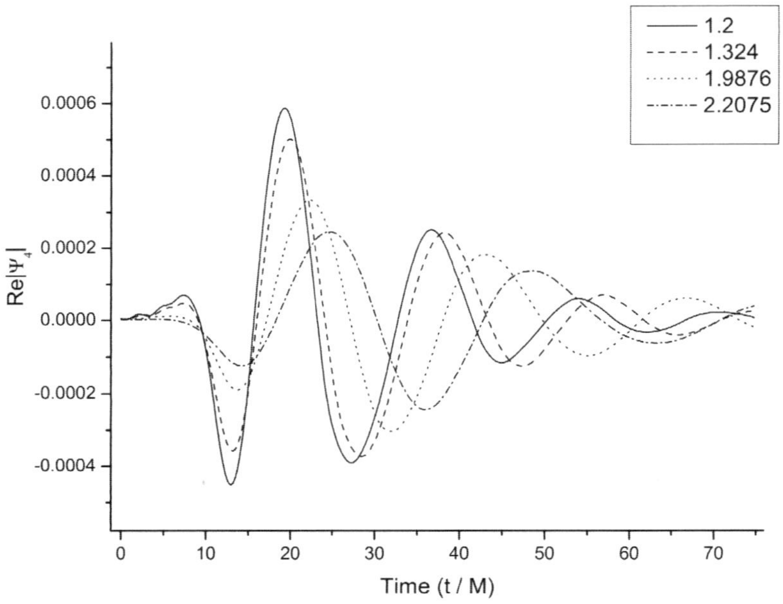

Figure 5. Gravitational waves used to test the neural networks, with some variations in mass ratio. The label of each curve indicates the mass ratio between the black holes.

the order of the same inputs as well as the iterations. The mean square error (MSE) is used to measure the performance of neural networks:

$$MSE = \sum_i \frac{(\tilde{m}^i - m^i)^2}{N} \tag{24}$$

where $\tilde{m}$ and m are the real and the estimated value with index i and N is the number of mass ratios to be predicted. Since several neural networks are generated with a determined number of hidden neurons, an average of this MSE is calculated as shown in table 1.

Despite the number of samples, the network has reached a considerable low error in the validation test 1. Table 2 displays the averaged percent error predictions for each real mass ratio. A more considerable difference is observed in the region around 2.0 since there is a large gap in the training samples from 1.7 and 2.1 mass ratios.

5. Final comments

Using a neural network we have predicted the mass ratio of binary black hole out of the gravitational wave form produced during the collision, with a success of less then 1% in the best case and around 10% in the worst example, this depends on both the number of training patterns and whether the signal to be interpreted lies far from a sampled region.

Neural networks performance		
Neural network configuration	Training averaged MSE	Test averaged MSE
NN-200	1.12436E-03	2.0302E-02
NN-300	1.48354E-04	8.5020E-03
NN-400	1.1737E-03	8.4419E-03
NN-500	1.2051E-03	1.1810E-02
NN-600	1.1591E-03	2.0541E-02
NN-700	1.5617E-03	1.9152E-02
NN-800	1.4744E-03	1.4716E-02

Table 1. An average of the mean square error eq. 24, is calculated over all training and validation samples, for each initialized random neural network configuration with 500 trainings. The NN-XXX indicates the number of hidden neurons.

	Real mass ratio			
Neural network	1.2	1.324	1.9876	2.2075
NN-200	10.992	1.700	12.551	0.75
NN-300	3.169	0.800	8.846	0.158
NN-400	5.109	1.792	8.540	0.226
NN-500	6.425	2.123	9.300	0.271
NN-600	11.904	6.647	12.046	0.2548
NN-700	11.079	2.826	11.641	7.970
NN-800	5.459	1.364	11.386	0.310

Table 2. Averaged percent mean squared error of the outcomes versus the real mass ratios, thrown by the 4 neural networks with different architectures. The NN-XXX indicates the number of hidden neurons.

We plan to extend this model to the analysis of more parameters concerning binary black holes, and eventually incorporate the algorithm to signal filtering libraries that will allow to measure the effectiveness in real situations of our methods. In regard to neural network improvements, it is needed to increase the number of training waves, explore a larger space of neural network architectures, compare with other training methods or even make use of genetic algorithms to minimize error. As well as define new ways to input information into the neural network like spectral analysis that potentially will speed up any parameter estimate.

Acknowledgments

This research is partly supported by CIC-UMSNH projects 4.9 and 4.23.

References

[1] B.S. Sathyaprakash and Bernard F. Schutz, Living Rev. Relativity 12 (2009), 2
[2] Gregory M. Harry and LIGO Scientific Collaboration, Advanced LIGO: The next generation of gravitational wave detectors, Classical Quantum Gravity 27, 084006 (2010).
[3] J. Abadie et al. (LIGO Scientific Collaboration and Virgo Collaboration), Classical Quantum Gravity 27, 173001 (2010).
[4] V. Raymond, M. V. van der Sluys, I. Mandel, V. Kalogera, C Rover and N. Christensen, Classsical Quantum Gravity 26, 114007 (2009).
[5] R. J. E. Smith, K. Cannon, C. Hanna, D. Keppel, and I. Mandel, Phys. Rev. D 87, 122002 (2013).
[6] http://einsteintoolkit.org/
[7] Stuart J. Russell and Peter Norvig, Artificial Intelligence: A Modern Approach, (Prentice-Hall, NJ,1995), pp. 563-566

[8] W. McCulloch and W. Pitts, Bull. of Math. Biophys. 5, (1943), 115.
[9] B. Karlik and V. Olgac. Int. J. of Artificial Intelligence and Expert Syst., 1(4), (2010), 111.
[10] Frank Rosenblatt, Perceptrons and the Theory of Brain Mechanisms, (Spartan, NY, 1992).
[11] G. Cybenko, Mathematics of Control, Signals and Systems, 2(4),(1989), 599.
[12] Martin Møller, Neural Networks Lecture Notes in Computer Science, 6(4), (2012), pp. 525-533.
[13] IOP Publishing is to grateful Mark A Caprio, Center for Theoretical Physics, Yale University, for permission to include the `iopart-num` BIBTEXpackage (version 2.0, December 21, 2006) with this documentation. Updates and new releases of `iopart-num` can be found on `www.ctan.org` (CTAN).
[14] G. B Cook, Living Rev. Relativity 3 (2000), 5.

Thermal properties for an ensemble of polymer Fermi oscillators

Guillermo Chacón-Acosta
Applied Mathematics and Systems Departament, Universidad Autónoma Metropolitana-Cuajimalpa, Av. Vasco de Quiroga 4871, México D. F. 05348, Mexico
E-mail: gchacon@correo.cua.uam.mx

Angel A. García-Chung
Physics Department, Universidad Autónoma Metropolitana-Iztapalapa, San Rafael Atlixco 186, México D. F. 09340, Mexico
E-mail: alechung@xanum.uam.mx

Héctor H. Hernandez-Hernandez
Engineering Faculty, Nuevo Campus Universitario Universidad Autónoma de Chihuahua, Chihuahua 31125, Mexico
E-mail: hhernandez@uach.mx

Abstract.
Polymer quantum mechanics is a model inspired on loop quantum gravity in which one can study, in a simplified way, some properties of certain quantum mechanical models. There is a length parameter in this model, known as the polymer scale, comprising the discreteness introduced in this particular quantization. There is a recent analysis on quantum fields where the Fermi oscillator is *polymerized* by means of a non-analytic representation of the corresponding Weyl super-algebra, its energy spectrum acquires modifications by the introduction of polymer parameters, which turn out to be super-numbers. In this work we present the first step in studying thermostatistical properties of an ensemble of Fermi oscillators. As an initial approximation we consider the polymer parameters as real deviations from their usual values. We obtain modifications to the thermal properties of the system in terms of polymer parameters. In the last section we critically discuss the possible physical significance of the results.

1. Introduction

The scheme of Loop Quantum Gravity (LQG) offers a diffeomorphism invariant and non perturbative quantum representation of gravity [1] at the kinematical level. This scheme, originally applied purely to the gravitational sector, has been extended to include matter fields like the scalar, gauge and fermion fields [2, 3]. Due to the intricate mathematical apparatus of the full LQG, a toy model, called "Polymer quantum mechanics" (PQM) [4], has served as a guide to bypass aspects related to the infinite degrees of freedom within the field theory, and to single out some of its aspects, specially as a system with a finite number of degrees of freedom.

Among the systems that PQM has been able to analyze are the study of the problematics to incorporate a Galilean symmetry to it [5], an analysis of singular potentials [6, 7], a particle in the half real line [8], the time diffraction [9] and the tunnel effect of polymer particles [10].

The statistical mechanics of systems built up by a bunch of particles has also been studied in [11, 12]. In that work corrections to the thermodynamics were found, induced by the so called polymer length scale, a parameter arisen in the polymer quantization of bosonic fields [4]. Some other thermal systems, such as hydrogenoid gases [13], have been analyzed. Particularly the study of a polymer Bose-Einstein condensate gives us bounds in the polymer length [14]; it was also possible to obtain modification in the mass of a compact star due to polymer effects by studying a completely degenerate Fermi gas at zero temperature [15, 16].

There have also been attempts to study field theory in this context, resulting in what is known as Polymer Quantum Field Theory (PQFT). The idea of replacing the tower of harmonic oscillators in which the field can be decomposed by polymer oscillators was advanced in [17]. The result is that the propagator has Planck-scale corrections. When the energy of the tower of polymer harmonic oscillators is close to the Planck energy, the propagator departs considerably of the standard propagator and admits tachyonic mass poles. The standard propagator is obtained as a low energy limit. In the same vein a study of the high energy behavior of the Dirac field in the polymer framework has been recently proposed. The idea was to apply PQM to each of the Fermi oscillators into which the Dirac fiel can be decomposed [18]. The result can be found in [19], where an unconventional modification on the free propagator for the Dirac field, which involves Grassmann-type quantities, was obtained.

These results demand a deeper analysis, not only for the transition amplitudes derived in the coupling to other matter fields like the $U(1)$ field within the standard model, but also for their statistical implications and their perspective into phenomenological modeling. This exploration is the main intent of this contribution.

In the first section, we briefly review the main aspects of the PQM for the Fermi oscillator. In the second section we present the thermal properties of an ensamble of polymer Fermi oscillators that is essentially a gas built up of two-level polymer particles. In the third section a discussion of the possibilities of interpreting the super-polymer parameters using the statistical results is presented. As a first exploration of their meaning, we treat them as real perturbations, near to their standard real value without polymer corrections. This was done following previous work on phenomenological models [11]-[16]. The main difference is that in those models the polymer parameter was easily interpretable as a fundamental length scale, while in the fermionic case it cannot be done. This test, however, gives us an idea of the possible effects that the polymer corrections may have on the thermal properties of a simple statistical system. This work is not intended to be a definitive analysis, as we mentioned above it is a first approach aiming to have a point of comparison with the corresponding super-corrections; subsequent considerations will be performed on a later work in preparation.

2. Polymer Fermi oscillator

2.1. The Fermi oscillator

The Fermi oscillator [20] is a mechanical system whose configuration space is $\mathbb{R}^2_a$, the two dimensional real anticommutative space. The Dirac formalism for constrained systems [21] can be applied to it, yielding a quantum model with a Hamiltonian of the form

$$\widehat{H} = \frac{\omega}{2i}(\widehat{x}_1\widehat{x}_2 - \widehat{x}_2\widehat{x}_1). \tag{1}$$

The corresponding canonical anticommutation relations (CAR) take the form

$$[\widehat{x}_i, x_j]_+ = \widehat{x}_i\widehat{x}_j + \widehat{x}_j\widehat{x}_i = \delta_{ij}\hbar\widehat{1}, \tag{2}$$

where $\widehat{1}$ and $\widehat{x}_i$ (with $i = \{1, 2\}$) denote the generators of the quantum algebra. An irreducible representation of this algebra can be realized on a super Hilbert space to allow multiplication by super numbers. In our case, we consider the super Hilbert space $\mathcal{H}_s$ formed by superanalytic functions [20]. A vector state $\Psi \in \mathcal{H}_s$ is a superfunction $\Psi : I\!R_a \to \Lambda_\infty; x \mapsto \Psi(x)$. Here Λ_∞ is the space of supernumbers [22]. The representation of the CAR is thus of the form

$$\widehat{x}_1\Psi(x) = \frac{\sqrt{2\hbar}}{2}\Psi(x)\left(x + \frac{\mathrm{d}^R}{\mathrm{d}x}\right), \qquad \widehat{x}_2\Psi(x) = \frac{\sqrt{2\hbar}}{2i}\Psi(x)\left(x - \frac{\mathrm{d}^R}{\mathrm{d}x}\right), \tag{3}$$

where $\frac{\mathrm{d}^R}{\mathrm{d}x}$ is the right derivative. Because any superfunction in $\mathcal{H}_s$ may be decomposed as $\Psi(x) = \Psi_0 + \Psi_1 x$, where $\Psi_0 = \Psi(x = 0)$ and $\Psi_1 = \left[\Psi(x)\frac{\mathrm{d}^R}{\mathrm{d}x}\right]\Big|_{x=0}$, the stationary Schrödinger equation (1) takes an algebraic form

$$\widehat{H}\Psi(x) = -\frac{\hbar\omega}{2}(\Psi_0 - \Psi_1 x) = E\Psi(x). \tag{4}$$

There are only two eigenvalues for (4) which are $E_\pm = \pm\frac{\hbar\omega}{2}$ and their respective basis of eigenfunctions is $\Psi_-(x) = 1$ and $\Psi_+(x) = x$. It is worth mentioning that the advantages of a representation on a super Hilbert space is that it allows multiplication (by the left or the right) of supernumbers, as well as for the algebra. Notice that, in this case, the energy spectrum is real and corresponds to a two level system. This does not happen in the polymer case, so the interpretation of the polymer superparameters is not as direct as in this case.

2.2. Polymerization of the Fermi oscillator

Let us review the main ideas in reaching the polymer version of this mechanical system which, basically, resembles the construction of the polymer version of the harmonic oscillator [23, 24]. The first step is to construct the *super* version of the Weyl algebra, which can be performed using the standard representation (3) as follows

$$\widehat{W}(\theta_1, \theta_2)\Psi(x) := e^{\theta_1\widehat{x}_1 - i\theta_2\cdot\widehat{x}_2}\Psi(x) := [1 + \theta_1\cdot\widehat{x}_1 - i\theta_2\cdot\widehat{x}_2 - i\theta_1\theta_2\cdot\widehat{x}_1\widehat{x}_2]\Psi(x). \tag{5}$$

where the parameters θ_1 and θ_2 are real anticommutative supernumbers. Using (3) and (5), the product of two generators $\widehat{W}$ is well defined and takes the form

$$\widehat{W}(\theta_1, \theta_2)\cdot\widehat{W}(\theta_1', \theta_2') = e^{\frac{\hbar}{2}(\theta_1\theta_1' + \theta_2\theta_2')}\widehat{W}(\theta_1 + \theta_1', \theta_2 + \theta_2'). \tag{6}$$

Equation (5) defines a regular representation of the Weyl superalgebra whose multiplication is given by (6). Its polymer version is just a nonregular, i.e. non super analytic representation of this superalgebra [23]. To get such a representation we discretize the real anticommutative space and then define an abstract super vector space basis $|x_j\rangle$. The linear combinations

$$|\Psi\rangle = \sum_j^N |x_j\rangle\Psi_{x_j}. \tag{7}$$

are the super vectors which span the super vector space $\mathcal{F}_p$ after giving it a vector superspace structure [23].

Super Hilbert spaces do not have an inner product structure but rather a bijective map between the vector superspace and its dual [20]. The dual space $\mathcal{F}_p^*$ is formed by elements

$$\langle\Psi| = \sum_j^N \Psi_{x_j}\langle x_j|. \tag{8}$$

In the polymer case the bijection is constructed as

$$\langle x|x'\rangle = \delta_{x,x'}, \tag{9}$$

where $\delta_{x,x'}$ is the Kronecker delta. Then $\mathcal{H}_p$ denote the super Hilbert space formed with $\mathcal{F}_p$ together with the bijection defined above, where a representation of the Weyl superalgebra generators can be realized as

$$\widehat{W}(\theta_1,\theta_2)|x\rangle := e^{\frac{\hbar}{2}\theta_1\theta_2}e^{\frac{\sqrt{2\hbar}}{2}(\theta_1-\theta_2)x}\left|x+\frac{\sqrt{2\hbar}}{2}(\theta_1+\theta_2)\right\rangle. \tag{10}$$

Incorporating (9) with (10) makes the Weyl superalgebra non regular in the sense that superanalyticity is not available and hence we cannot recover the generators of the Clifford superalgebra (3).

Finally one can analyze the dynamics of this system. Because the Hamiltonian (1) is not directly related to the Weyl superalgebra generators (6), due to the lack of superanalyticity, a Hamiltonian that can be expressed in terms of Weyl generators is proposed in [23]

$$\widehat{H} \to \widehat{H}_p = \frac{\hbar\omega}{2}\left[2-\widehat{W}(\theta_1,\theta_2)-\widehat{W}(-\theta_1,-\theta_2)\right]. \tag{11}$$

Notice its similarity with the bosonic case [4].

We now proceed to solve the Schrödinger equation $\widehat{H}_p\Psi = E\Psi$, but first we introduce a *momentum representation* and a *Fourier transform*, as in the bosonic case

$$\langle\Psi|k\rangle := \tilde{\Psi}(k) = \sum_{x_j}\Psi_{x_j}\langle x_j|k\rangle, \quad \text{where} \quad \langle k|x\rangle := e^{-kx} = \overline{\langle x|k\rangle}. \tag{12}$$

Let us denote by $\mathcal{H}_{poly}$, rather than $\mathcal{H}_p$, the Hilbert space with the structure

$$\mathcal{H}_{poly} = \bigoplus_{x_0}\mathcal{H}^{x_0}_{poly}. \tag{13}$$

Any $|\Psi_{x_0}\rangle \in \mathcal{H}^{x_0}_{poly}$ takes the form

$$|\Psi_{x_0}\rangle = \sum_j |x_0 + \frac{\sqrt{2\hbar}}{2}j(\theta_1-\theta_2)\rangle\Psi^j_{x_0}. \tag{14}$$

The Schrödinger equation for $x_0 = 0$ reads

$$\Psi_0^{j-1}e^{-\frac{\hbar}{2}\theta_1\theta_2}e^{-\hbar(j-1)\theta_1\theta_2} + \Psi_0^{j+1}e^{-\frac{\hbar}{2}\theta_1\theta_2}e^{+\hbar(j+1)\theta_1\theta_2} - 2\left(1-\frac{E\theta_1\theta_2}{\omega}\right)\Psi_0^j = 0, \tag{15}$$

where E denotes the energy eigenvalue. Let us now perform a Fourier transformation to go to momentum representation using (12). To do so we multiply by $e^{\frac{\sqrt{2\hbar}}{2}jk(\theta_1-\theta_2)}$ and sum over j. This gives us finally

$$\left[\frac{\hbar\omega}{2}-E\right]\theta_1\theta_2\tilde{\Psi}_0 - \left[\frac{\hbar\omega}{2}+E\right]\theta_1\theta_2\tilde{\Psi}_1 k = 0. \tag{16}$$

Here, $\tilde{\Psi}_0 = \tilde{\Psi}(k=0)$ and $\tilde{\Psi}_1 = \tilde{\Psi}(k)\frac{\mathrm{d}^R}{\mathrm{d}k}\Big|_{k=0}$ are the coefficients of the wave function in *momentum* representation. By using the following property of the wave functions

$$\sum_x \Psi_x e^{(a+k)x} = \tilde{\Psi}(k) + \left(\frac{\mathrm{d}^R}{\mathrm{d}k}\tilde{\Psi}(k)\right)a, \tag{17}$$

we obtain that eq. (16) possesses the eigenvalues

$$E_{\pm} = \pm\frac{\hbar\omega}{2} + (e_{\pm})\hbar\omega(\hbar\theta_1\theta_2). \tag{18}$$

In this expression, $e_{\pm}$ are two commutative dimensionless arbitrary parameters. When we fix $e_{\pm} = 0$, $\theta_1, \theta_2 = 0$ we obtain the standard energy spectrum of the Fermi oscillator.

3. Thermal properties of an ensamble of polymer Fermi oscillators

Considering the energy spectrum (18) we can immediately calculate the canonical partition function under the assumption that the statistics does not change [12]. As there are only two possible values for the energy, Z becomes

$$Z = \exp\left(-\frac{E_+}{kT}\right) + \exp\left(-\frac{E_-}{kT}\right) = \exp\left(-\frac{\hbar\omega}{2kT}\Delta_+\right) + \exp\left(\frac{\hbar\omega}{2kT}\Delta_-\right), \tag{19}$$

or in terms of hyperbolic functions, as is commonly written

$$Z = \cosh\left(\frac{\hbar\omega}{2kT}\Delta_+\right) - \sinh\left(\frac{\hbar\omega}{2kT}\Delta_+\right) + \cosh\left(\frac{\hbar\omega}{2kT}\Delta_-\right) + \sinh\left(\frac{\hbar\omega}{2kT}\Delta_-\right). \tag{20}$$

In all previous expressions k is the Boltzmann constant, T is the gas temperature and the following quantities are defined

$$\Delta_+ := 1 + 2\hbar e_+\theta_1\theta_2, \quad \Delta_- := 1 - 2\hbar e_-\theta_1\theta_2, \tag{21}$$

comprising all the polymer effects discussed in the previous section. Also notice that, by making $\Delta_+ = \Delta_- = 1$, we recover the usual partition function for a two level system (as a magnetic dipole) $Z = 2\cosh\left(\frac{\hbar\omega}{2kT}\right)$,[25].

Just as in [11], we can compute all the thermodynamic functions recalling that the partition function for a system of N particles is Z^N. Thus, the free energy is

$$F = -NkT\ln Z = -NkT\ln\left[e^{-\frac{\hbar\omega}{2kT}\Delta_+} + e^{\frac{\hbar\omega}{2kT}\Delta_-}\right]. \tag{22}$$

The internal energy and heat capacity are as follows

$$U = NkT^2\frac{\partial Z}{\partial T} = \frac{N\,\hbar\omega}{2}\left[\frac{\Delta_+\exp\left(-\frac{\hbar\omega}{2kT}\Delta_+\right) - \Delta_-\exp\left(\frac{\hbar\omega}{2kT}\Delta_-\right)}{\exp\left(-\frac{\hbar\omega}{2kT}\Delta_+\right) + \exp\left(\frac{\hbar\omega}{2kT}\Delta_-\right)}\right], \tag{23}$$

$$C_V = \left(\frac{\partial U}{\partial T}\right)_V = \frac{N\,\hbar^2\omega^2}{16kT^2}(\Delta_+ + \Delta_-)^2\,\mathrm{sech}^2\left[\frac{\hbar\omega}{4kT}(\Delta_+ + \Delta_-)\right]. \tag{24}$$

The entropy can be obtained directly by subtracting the free energy (22) to the internal energy:

$$S = \frac{U-F}{T} = \frac{N\,\hbar\omega}{2T}\left[\frac{\Delta_+\,e^{-\frac{\hbar\omega\Delta_+}{2kT}} - \Delta_-\,e^{\frac{\hbar\omega\Delta_-}{2kT}}}{e^{-\frac{\hbar\omega\Delta_+}{2kT}} + e^{\frac{\hbar\omega\Delta_-}{2kT}}}\right] + Nk\ln\left[e^{-\frac{\hbar\omega\Delta_+}{2kT}} + e^{\frac{\hbar\omega\Delta_-}{2kT}}\right]. \tag{25}$$

Although expressions (22)-(25) give us the form of the thermodynamic quantities of an ensemble of Fermi oscillators, it is still necessary to have a correct interpretation of the parameters $\Delta_{\pm}$ since, as pointed out in the previous section, these are super-parameters. However, as a first approximation for this analysis, we suppose $\Delta_{\pm}$ behaves as a perturbative quantity. The behavior of the thermodynamic variables under this assumption is shown in Figures 1-4. Suppose the effect of the polymer parameters is to reduce or slightly increase the real non-polymer value, so what we see in the figures is a graphical comparison of the functions (22)-(25) for values of $\Delta_\pm$ close to one, for which we recover the values of the functions without polymer effects.

At low temperature the system is in the lower energy level and its internal energy U is $-\hbar\omega\Delta_-/2$, this appears shifted due to polymer effects. At high temperatures both energy states are equally likely to be occupied, so that the internal energy tends to a minimum value $\hbar\omega(\Delta_+ - \Delta_-)/2$, Fig. 1. The polymer effect is to spread the energy values into a band centered in the non-polymer value, so that, at high temperatures, the internal energy has a nonzero value that depends exclusively on the polymer effects present. Note that the same thing occurs with the free energy due to its functional form, Fig. 2.

At $T = 0$, the entropy is exactly zero, even with polymer modifications, which cancel out each other. The entropy increases with temperature reflecting the freedom that the system acquires in different states. Amusingly, the entropy tends to $k \ln 2$ exactly, as usual for high temperature, even with polymer corrections. Certainly it seems that the polymer effect on the entropy is to change the way it tends towards the usual limits, provided that $\Delta_+ \neq \Delta_-$, Fig. 3.

Something similar occurs with the heat capacity. The main difference is that the heat capacity tends to zero at both low and high temperature limits. Actually, C_V has a maximum at a temperature approximately equal to $T_c \approx \frac{\hbar\omega}{4k}(\Delta_+ + \Delta_-)$, to be precise at the roots of $X - \tanh\frac{1}{X} = 0$, with $X = \frac{4kT}{\hbar\omega(\Delta_+ + \Delta_-)}$. This maximum, known as the Schottky anomaly, separates two behaviors of the system, namely when only the lowest energy state is occupied and when both levels are occupied. Notice that the polymer effect when $\Delta_+ \neq \Delta_-$ is to shift the critical temperature, keeping the same value for the maximum. This behavior can be seen in Fig. 4.

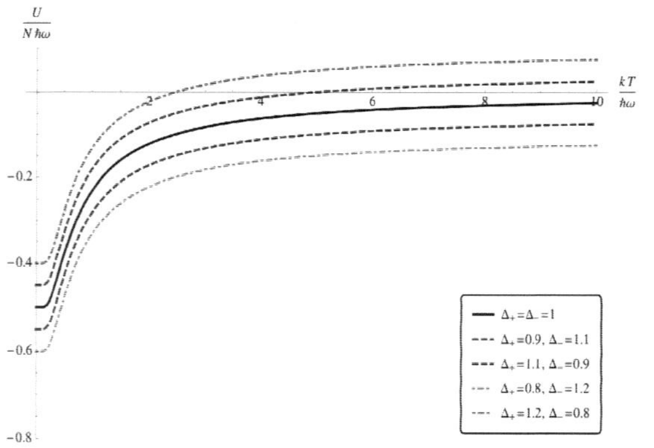

Figure 1. Internal energy U with polymer corrections. Black solid curve is the usual value with $\Delta_\pm = 1$. Blue and purple (dashed) curves have values close to one (from below and above respectively). Red and orange (dot-dashed) curves have values a little bit far from 1.

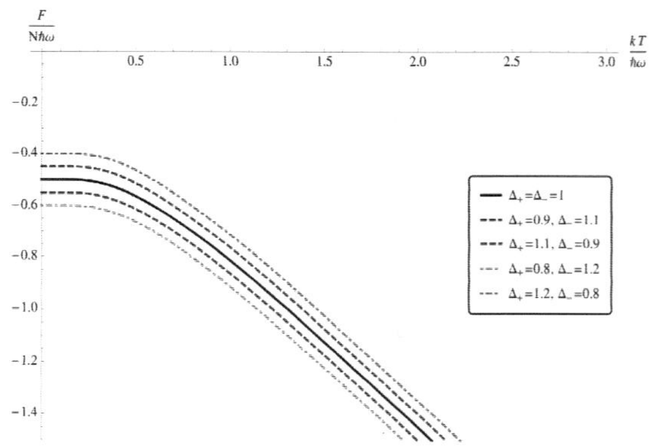

Figure 2. Free energy F with polymer corrections. Black solid curve is the usual behavior with with $\Delta_\pm = 1$. Blue and purple (dashed) curves have values close to one (from below and above respectively). Red and orange (dot-dashed) curves have Δ values deviating slightly more from 1.

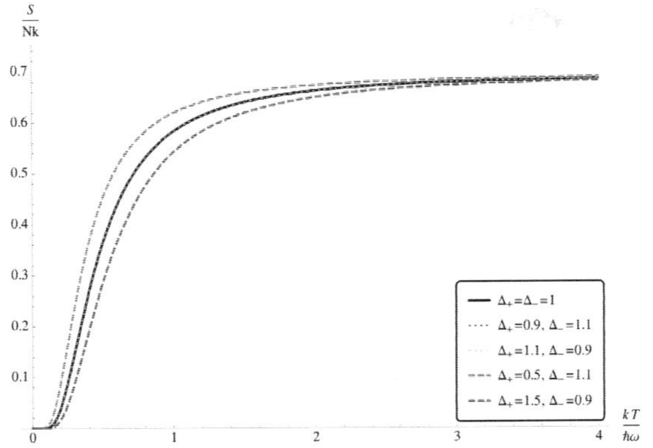

Figure 3. Entropy S with polymer corrections. Black solid curve is the usual behavior for $\Delta_{\pm} = 1$. Blue and green (dotted) curves have values close to one (from below and above respectively), but maintaining the same sum. Red and orange (dashed) curves have $\Delta_{+} \neq \Delta_{-}$. Note that the behavior of the entropy changes depending on the value of the polymer corrections, although its limits for high and low temperatures are the same.

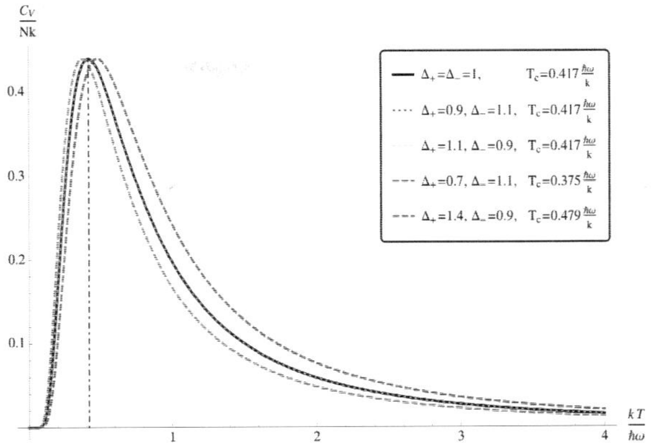

Figure 4. Heat capacity C_V with polymer corrections. Black solid curve is the usual behavior for $\Delta_{\pm} = 1$. Blue and green (dotted) curves have values close to one (from below and above respectively), but with the same sum. Red and orange (dashed) curves have $\Delta_{+} \neq \Delta_{-}$. The behavior of C_V is modified by the polymer corrections, although it tends to the same values for high and low temperatures. The dot-dashed vertical line indicates the critical temperature T_c with no corrections. Notice that in the case $\Delta_{+} \neq \Delta_{-}$, T_c is shifted with respect to that value.

4. Discussion

In this work we studied a simplified model for the thermostatistical properties of an ensemble of polymer Fermi oscillators. Thermal variables are modified due to polymer terms. As a first guess for the polymer parameters we treat them perturbatively near to the non-polymer real value. Under such assumption we found that polymer modification for energy variables, such as U and F, make them to slightly vary from the non corrected value. For low temperatures, it modifies the minimum value of the internal energy while for high temperature makes U reach a nonzero value.

On the other hand we observe for the entropy and heat capacity that the behavior at low and high temperatures is not affected by the polymer corrections. However, in the special case where $\Delta_{+} \neq \Delta_{-}$ they exhibit a change in its behavior at an intermediate temperature. Indeed, the critical temperature at which the C_V exhibits a maximum is shifted by polymer effects.

It should be emphasized that this discussion is made under the assumption that polymer effects, encoded in the quantities $\Delta_{\pm}$, are real quantities that slightly vary from the usual non-polymer value. This assumption is very stringent in light of what we discussed in the first section, where it was found that the polymer parameters are actually super-parameters. Eq. (19) and expressions derived thereforth are actually formal constructs with no *a priori* physical meaning. For example, the logarithm of a supernumber must be defined in terms of a power series: it is in that sense that one must treat other functions. Because the super-corrections were replaced with real perturbations, the figures shown may not necessarily have the complete physical content, however, they serve as a reference for when polymer modifications are taken as actual super-numbers. In this sense, the first thing one need to do is write the partition function (19) in terms of Grassmanian numbers and to establish which additional criteria is needed to consider corrections.

It might happen that in some special representation the influence of the spectrum (18) in phenomenology is not straightforward. Certainly, we may think in a system whose microscopic variables can be accurately measured, so one can bound and understand their effects. Indeed, recall that θ_1 and θ_2 appears when we write the Weyl super algebra in the fermionic case. These parameters are equivalent to the polymer length in the bosonic case, which is fully identifiable with a length scale; but not so in this case. Moreover, the $e_{\pm}$ parameter appears when solving equation (16) as another polymer super-parameter. It is possible that both parameters can be fixed when the polymer Dirac field is coupled to other matter fields. The main difficulty in the present construction is that there is not a clear definition of a Grassmanian observable. For example, the definition of a physical observable given in [20], does not prevent or limit the Grassmannian contribution to the thermodynamic quantities, so a reinterpretation of the parameters is required that could occur through a generalization of the concept of observable taking into account the singular representations appearing in loop quantum gravity and polymer quantum mechanics [24].

5. Acknowledgements

Partial support from the grant CONACYT CB-2008-01/101774 is acknowledged.

References

[1] A. Ashtekar, Phys. Rev. D **36**, 1587 (1987).
[2] C. Rovelli, *Quantum Gravity*, (Cambridge University Press, Cambridge, England, 2004).
[3] T. Thiemann, *Introduction to Modern Canonical Quantum General Relativity*, (Cambridge University Press, Cambridge, England, 2007).
[4] A. Ashtekar, S. Fairhurst and J. L. Willis, Class. Quant. Grav. **20**, 1031 (2003).
[5] W. D. Chiou, Classical Quantum Gravity **24** (2007), 2603 2620.
[6] V. Husain, J. Louko and O. Winkler, Phys. Rev. D **76** (2007), 084002.
[7] G. Kunstatter, J. Louko and J. Ziprick, Phys. Rev. A **79** (2009), 032104.
[8] G. Kunstatter and J. Louko, J. Phys. A **45**, 305302 (2012).
[9] A. Martín-Ruiz, Phys. Rev. D **90**, 125027 (2014).
[10] D. A. Demir, O. Sargın, Phys. Lett A **378**, 3237 (2014).
[11] G. Chacón-Acosta, E. Manrique, L. Dagdug and H. A. Morales-Técotl, SIGMA **7**, 110 (2011).
[12] G. Chacón-Acosta, L. Dagdug and H. Morales-Técotl, AIP Conf. Proc. **1396**, 99-103 (2011).
[13] G. Chacón-Acosta, AIP Conf. Proc. **1473**, 153-162 (2012).
[14] E. Castellanos and G. Chacón-Acosta, Phys. Lett. B **722**, 119 (2013).
[15] G. Chacón-Acosta and H. H. Hernández, AIP Conf. Proc. **1548**, 179-182 (2013).
[16] G. Chacón-Acosta and H. H. Hernandez-Hernandez. Int. J. Mod. Phys. D **24**, 1550033 (2015).
[17] G. M. Hossain, V. Husain and S. S. Seahra, Phys. Rev. D **82**, 124032 (2010).
[18] A. A. García-Chung and H. A. Morales-Técotl, AIP Conf. Proc. **1473**, 163 (2011).
[19] A. A. García-Chung and H. A. Morales-Técotl, Phys. Rev. D **89**, 065014, (2014).
[20] B. S. DeWitt, *Supermanifolds*, (Cambridge University Press, Cambridge, England, 1984) 1992.
[21] M. Henneaux and C. Teitelboim, *Quantization of Gauge Systems*, (Princeton University Press 1991)
[22] A. Khrennikov, *Superanalysis*, (Kluwer Academic Publishers 1999)
[23] A. A. García-Chung, H. A. Morales-Técotl and J. D. Reyes, AIP Conf. Proc. **1548**, 161 (2013).
[24] A. A. García-Chung, H. A. Morales-Técotl and J. D. Reyes, "On the GNS formalism for polymer Fermi oscillator" (In preparation).
[25] W. Greiner, Ludwig Neise, H. Stöcker, *Thermodynamics and Statistical Mechanics*, (Springer 1995).

An Effective Non-Commutative Loop Quantum Cosmology

Abraham Espinoza-García, M. Sabido, J. Socorro
Departamento de Física de la DCeI de la Universidad de Guanajuato-Campus León
C.P. 37150, Guanajuato, México
E-mail: abraham@fisica.ugto.mx
E-mail: msabido@fisica.ugto.mx
E-mail: socorro@fisica.ugto.mx

Abstract. We construct a Non-Commutative extension of the Loop Quantum Cosmology effective scheme for the open FLRW model. We start from the holonomized Hamiltonian and implement a canonical non-commutativity among the matter degree of freedom and the holonomy variable, in the volume representation. We obtain a noncommutativie extension of the modified Friedmann equation which arises in Loop Quantum Cosmology for a particular case of the theta deformation.

1. Introduction

The idea of Noncommutativity in spacetime is quite old (1947) [1]. It was proposed as an attempt to regularize Quantum Field Theory before the renormalization program was established. Due to its non-local behavior, Noncommutativity was quickly forgotten after renormalization proved to be successful. In the 1970's M. Flato and co-workers proposed an alternative path to quantization [2], in which a deformation of the Poisson structure of classical phase space is performed and is encoded in the moyal $\star$-product [3] and generalizations of it (for a review see [4]). In the early 1980's mathematicians led by A. Connes succeeded in formulating what they called Noncommutative Geometry [5], motivated by generalizating a classic theorem characterizing C^*-algebras.

Recently, the noncommutative paradigm has resurrected, mainly due to results in String Theory [6, 7], in which Yang-Mills theories in a noncommutative space arise in different circumstances as effective theories when taking certain simple limits (for instance, the low energy limit). This renewed interest has led to a deeper understanding, from the physical and mathematical points of view, of noncommutative field theory (for a review see [8]). It is believed that in a full quantum theory of gravity the continuum picture of spacetime would no longer be consistent at distances comparable to the Planck length $\ell_p \sim 10^{-35}$cm, a quantization of spacetime itself could be in order. Furtheremore, since quite ago, it has been argued [9] that measurements of position can not be performed to better accuracies than the Planck length, since spacetime itself would be modified due to the energy required to peform such measurments.

A possible way to model this effects could be via an uncertainty relation for the spacetime coordinates of the form

$$[x^i, x^j] = i\theta^{ij} \quad (1)$$

This commutation relation is the starting point of noncommutative field theory. Attempts to implement this idea have led to different proposals for a noncommutative theory of gravity [10]. Different incarnations of noncommutative gravity have a common feature, they are highly non-linear theories, which makes them very difficult to work with.

A possible way to study noncommutativity effects in the early universe was proposed by Garcia-Compean et al [11]. They implemented noncommutativity in configuration space (in contrast to noncommutative spacetime), but only after a symmetry reduction of spacetime had been imposed and the Wheeler-DeWitt quantization had been carried, giving rise to a noncommutative quantum cosmology (mathematically similar to noncommutative quantum mechanics [12]). Later, G. D. Barbosa and N. Pinto-Neto [13] introduced this minisuperspace noncommutativity already at the classical level. The idea is that, perhaps, this effective noncommutativity could incorporate novel effects and insights of a full quantum theory of the gravitational field, alongside with providing a simple framework for studying the implications of such possible noncommutative effects in the (early) universe through cosmological models. Since the publication of these two seminal investigations, some works along this line have been conducted. For instance, the noncommutativity of the Friedmann-Robertson-Walker cosmology has been studied, as well as some of the Bianchi Class A models [14]. Quantum black holes have also been investigated within this framework [15].

On the other hand, Loop Quantum Gravity (LQG) [16, 17] is an attempt to quantize the gravitational field taking seriously the lessons from General Relativity, that is, it aims at a full (non-perturbative) background independent quantization of General Relativity. Loop Quantum Cosmology (LQC) [18, 19] is the quantization of cosmological (symmetry reduced) models following closely the ideas and methods of LQG. In this way, the LQC of the FLRW and some of the Bianchi Class A models in the presence of a massless scalar field (employed as internal time) has been constructed [20, 21, 22, 23, 24, 25, 26], in particular, as a result of the underlying quantum geometry, it has been shown that the loop quantization of the FLRW models features a bouncing which enables the resolution of the cosmological singularity [27]. The LQC of the inhomogeneous Gowdy model has also been constructed [28]. As a result of loop quantization, the Wheeler-DeWitt equation is no longer a differential equation, but a difference equation, which is difficult to work with even in the simplest models. In order to extract physics, effective equations based on a geometrical formulation of Quantum Mechanics [29] have been employed to study the consecuences of loop quantum corrections in cosmological models. For instance, the effective description of the FLRW models reproduces very well the behavior of the corresponding full loop quantization of such models. The present investigation aims at constructing a non-commutative effective scheme for the open FLRW model, in the pressence of a scalar field.

The manuscript is organized as follows: In section II we introduce the loop variables for the open FLRW model with a free standard scalar field and the corresponding effective scheme. In section III we recall the Weyl-Wigner-Moyal correspondence and the construction of a Noncommutative Quantum Mechanics based on this correspondence. Section IV is devoted to construct a non-commutative model for the effective Loop Quantum Cosmology of the open FLRW model resembling the Noncommutative Quantum Mechanics of section III, along the lines of references [11, 13].

2. Connection Variables and Effective Dynamics

Here we recall the formulation of the open FLRW and Bianchi I models in the Ashtekar-Barbero variables, with a free massless scalar field. The Ashtekar-Barbero variables cast General Relativity in the form of a gauge theory, in which phase space is described by a $\mathfrak{su}(2)$ gauge connection, the Ashtekar-Barbero connection A_a^i and its canonical conjugate momentum, the densitized triad E_i^a ($i, j, etc.$ denote internal $\mathfrak{su}(2)$ indices while $a, b, etc.$ denote spatial indices).

These quantities are defined as

$$A^i_a = \Gamma^i_a + \gamma K^i_a, \quad E^a_i = \sqrt{q} e^a_i \tag{2}$$

where $K^i_a = K_{ab} e^a_j$, with e^i_a and e^a_i the triad and co-triad, respectively ($e^i_a e^a_j = \delta^i_j$, $q_{ab} = \delta_{ij} e^i_a e^j_b$); Γ^i_a is defined through the spin connection ω^i_{aj} compatible with the triad ($\nabla_a e^i_b + \omega^i_{aj} e^j_b = 0$, ∇_a being the usual spatial covariant derivative) by the relation $\omega^i_{aj} = \epsilon_{ijk}\Gamma^i_a$ with ϵ_{ijk} the totally antisymmetric symbol; γ a real constant called the Barbero-Immirzi parameter. The canonical pair has the following Poisson structure

$$\{A^a_i(x), E^j_b(y)\} = 8\pi G\gamma \delta^j_i \delta^a_b \delta(x,y) \tag{3}$$

with $\delta(x,y)$ the Dirac delta distribution on the space-like hypersurface Σ. In these variables, the Hamiltonian constraint takes the form

$$C_{grav} = \left(\frac{1}{|E|}\epsilon_{ijk}\left[F^i_{ab} - (1+\gamma^2)\epsilon^{imn} K^m_a K^n_b\right] E^{aj} E^{bk}\right) \tag{4}$$

where $F^i_{ab} = \partial_a A^i_b - \partial_b A^i_a + \epsilon_{ijk} A^j_a A^k_b$ is the curvature of the Ashtekar-Barbero connection and E the determinant of the densitized triad. While the full gravitational Hamiltonian is

$$H_{grav} = \int d^3x N C_{grav} \tag{5}$$

2.1. Open FLRW model

for spatially flat homogeneous models the gravitational Hamiltonian reduces to [19]

$$H_{grav} = -\gamma^2 N \int_{\mathcal{V}} d^3x \frac{1}{|E|}\epsilon_{ijk} F^i_{ab} E^{aj} E^{bk} \tag{6}$$

where the integral is taken in a fiducial cell $\mathcal{V}$. When imposing also isotropy, the connection and triad can be described by parameters c and p, respectively, defined by [19]

$$A^i_a = cV_0\, {}^o e^i_a, \quad E^i_a = pV_0 \sqrt{{}^o q}\, {}^o e^a_i \tag{7}$$

where ${}^o q_{ab}$ is a fiducial flat metric (with which we can endow Σ due to the symmetries of the model) and ${}^o e^a_i$, ${}^o e^i_a$ are constant triad and co-triad compatible with ${}^o q_{ab}$; V_0 is the volume of $\mathcal{V}$ with respect to ${}^o q_{ab}$. These variables do not depend on the choice of the fiducial metric. The relation among these variables and the usual geometridynamical variables is

$$c = V_0^{1/3}\gamma\dot{a}, \quad p = V_0^{2/3} a^2 \tag{8}$$

We have the canonical relations

$$\{c, p\} = \frac{8\pi G\gamma}{3} \tag{9}$$

For convinience we perform the following change of variables

$$\beta = \frac{c}{\sqrt{p}}, \quad V = p^{\frac{3}{2}} \tag{10}$$

with the canonical relations

$$\{\beta, V\} = 4\pi G\gamma \tag{11}$$

In these variables the Hamiltonian (with a free massless scalar field as the matter content) takes the form ($N = 1$)

$$H = N\left(-\frac{3}{8\pi G\gamma^2}\beta^2 V + \frac{p_\phi^2}{2V}\right) \tag{12}$$

The Holonomy correction due to Quantum Gravity effects is coded in the replacement [19]

$$\beta \mapsto \frac{\sin\lambda\beta}{\lambda} \tag{13}$$

where $\lambda^2 = 4\sqrt{3}\pi\gamma\ell_p^2$ is the smallest eigenvalue of the area operator in the full Loop Quantum Gravity [17]. The resulting effective Hamiltonian is thus given by

$$H_{eff} = -\frac{3}{8\pi G\gamma^2\lambda^2}\sin^2(\lambda\beta)V + \frac{p_\phi^2}{2V} \tag{14}$$

The field equations are

$$\dot{\beta} = 4\pi G\gamma\frac{\partial H_{eff}}{\partial V} = -\frac{3}{\gamma\lambda}\sin^2(\lambda\beta) - 4\pi G\gamma\frac{p_\phi^2}{2V^2} \tag{15}$$

$$\dot{V} = -4\pi G\gamma\frac{\partial H_{eff}}{\partial\beta} = \frac{3}{\gamma\lambda}V\sin(\lambda\beta)\cos(\lambda\beta) \tag{16}$$

$$\dot{\phi} = \frac{\partial H_{eff}}{\partial p_\phi} = \frac{p_\phi}{V} \tag{17}$$

$$\dot{p}_\phi = -\frac{\partial H_{eff}}{\partial\phi} = 0 \tag{18}$$

Since $V = a^3$, $\frac{\dot{V}}{3V} = \frac{\dot{a}}{a} = \mathrm{H}$, where H is the Hubble parameter. Then, taking into account the effective Hamiltonian constraint, $H_{eff} = 0$, and the field equation for V we have,

$$\mathrm{H}^2 = \left(\frac{\dot{V}}{3V}\right)^2$$

$$= \frac{1}{\gamma^2\lambda^2}\left[\frac{8\pi G\gamma^2\lambda^2}{3}\frac{p_\phi^2}{2V^2}\left(1 - \frac{8\pi G\gamma^2\lambda^2}{3}\frac{p_\phi^2}{2V^2}\right)\right]$$

$$= \frac{8\pi G}{3}\rho\left(1 - \frac{\rho}{\rho_{max}}\right) \tag{19}$$

where $\rho = \frac{p_\phi^2}{2V^2} = \frac{3}{8\pi G\gamma^2\lambda^2}\sin^2(\lambda\beta)$ and ρ_{max} is the maximum value that the ρ can take in view of the effective Hamiltonian constraint, that is, $\rho_{max} = \frac{3}{8\pi G\gamma^2\lambda^2}$. The turning points of the volume function occur at $\beta = \pm\frac{\pi}{2\lambda}$, which correspond to a bounce. The last equality is the modified Friedmann equation, which incorporates holonomy corrections due to Loop Quantum Gravity. In the limit $\lambda \to 0$ (no area gap) we recover the ordinary Friedmann equation $\mathrm{H}^2 = \frac{8\pi G}{3}\rho$.

The relational evolution of V in terms of ϕ is given by

$$\frac{dV}{d\phi} = \frac{dV}{dt}\frac{dt}{d\phi} = \frac{3}{\gamma\lambda}\sin(\lambda\beta)\cos(\lambda\beta)\frac{V}{p_\phi} = \sqrt{12\pi G}V\left(1 - \frac{\rho}{\rho_{max}}\right)^{1/2} \tag{20}$$

where we have used the field equations for V and ϕ, and the effective Hamiltonian constraint.

3. Weyl-Wigner-Moyal Correspondence and Noncommutative Quantum Mechanics

The Weyl-Wigner-Moyal (WWM) correspondence [30] relates a phase space function f with its operator analog $\mathcal{W}(f) = \hat{f}$, it is based on the following map [31]

$$\hat{f}(\hat{q},\hat{p}) = \int \tilde{f}(\xi,\eta)e^{\frac{i}{\hbar}(\hat{p}_a\xi^a+\hat{q}^a\eta_a)}w(\xi,\eta)d^l\xi d^l\eta \tag{21}$$

where $\tilde{f}$ is the Fourier transform of f, w a weight function, and $dim(T^*Q) = 2l$; with $\hat{q}^a, \hat{p}_b$ satisfying the canonical Heisenberg algebra,

$$[\hat{q}^a,\hat{p}_b] = i\hbar\delta^a_b; \quad [\hat{q}^a,\hat{q}^b] = 0 = [\hat{p}_a,\hat{p}_b] \tag{22}$$

We define the $\star$-product with the help of this correspondence by

$$\mathcal{W}(f \star g) = \mathcal{W}(f)\cdot\mathcal{W}(g) \tag{23}$$

that is, $f \star g$ is the Weyl symbol of $\hat{f}\hat{g}$. For this case of the Heisenberg algebra (22) and in the symmetric ordering ($w = 1$) we have

$$f\star g = \exp\left(\frac{i\hbar}{2}P(f,g)\right) = fg + \sum_{r=1}^{\infty}\left(\frac{i\hbar}{2}\right)^r\frac{1}{r!}P^r(f,g) = f(x)\exp\left(\frac{i\hbar}{2}\partial_\mu\omega^{\mu\nu}\partial_\nu\right)g(x)$$
$$= fg + \frac{\hbar}{2}\{f,g\} + \mathcal{O}(\hbar^2) \tag{24}$$

where $P^r(f,g) = \omega^{\mu_1\nu_1}\cdots\omega^{\mu_r\nu_r}(\partial_{\mu_1\ldots\mu_r}f)(\partial_{\nu_1\ldots\nu_r}g)$ is the rth power of the Poisson bracket bidifferential operator, with $\omega^{\mu\nu}$ the components of the flat Poisson structure (**??**); x^μ are the phase space coordinates, denoted collectevely as x, the first n being the configuration coordinates q^a, the second n being the momenta p_a. This is the Moyal $\star$-product [3], it replaces the ordinary pointwise multiplication in the algebra of functions defined in phase space. The Moyal $\star$-bracket $f\star g - g\star f$ is thus responsible for realize the canonical Heisenberg algebra (22). Hence, (24) encodes a deformation of the classical phase space which yields the canonical Heisenberg algebra. This product is the cornerstone of deformation quantization [2, 4].

Now, consider the deformation of classical phase space which yields the algebra

$$[\hat{q}^a,\hat{p}_b] = i\hbar\delta^a_b; \quad [\hat{q}^a,\hat{q}^b] = i\theta\theta^{ab}; \quad [\hat{p}_a,\hat{p}_b] = i\kappa\kappa_{ab} \tag{25}$$

with θ^{ab}, κ_{ab} antisymmetric constat real matrices.

In light of the above prescription, we could encode this deformed Heisenberg algebra in a $\star$-product completely analogous to (24). The difference lies in the $\omega^{\mu\nu}$, which in this case would be of the form

$$i\begin{pmatrix} \theta\Theta & \hbar I \\ -\hbar I & \kappa \mathrm{K} \end{pmatrix} \tag{26}$$

This quantization leads us to a Noncommutative Quantum Mechanics, rather than the ordinary Quantum Mechanis obtained above.

4. Effective Non-Commutative Loop Quantum Cosmology

In view of the above discussion, we would like to consider effects of the following deformed Poisson algebra

$$\{\beta,\phi\} = \theta, \quad \{\beta, V\} = 4\pi G\gamma \tag{27}$$

Journal of Physics: Conference Series **654** (2015) 012003 doi:10.1088/1742-6596/654/1/012003

The above relations can be implemented by working with the shifted variables

$$\beta^{nc} = \beta + a\theta p_\phi, \quad \phi^{nc} = \phi + b\theta V \tag{28}$$

where a and b satisfy the relation $4\pi G\gamma b - a = 1$. In deed, we have the algebra

$$\{\beta^{nc}, \phi^{nc}\} = \theta, \quad \{\beta^{nc}, V\} = 4\pi G\gamma \tag{29}$$

Since we are not actually performing a deformation of the symplectic structure (the algebra among the basic phase space variables β, ϕ, V, p_ϕ is the same), the loop quantization of the original variables can be carried as usual, but with a slightly different Hamiltonian. Considering the steps above, we can therefore implement the effects of relation (28) in the effective Hamiltonian, which results in

$$H^{nc}_{eff} = -\frac{3}{8\pi G\gamma^2\lambda^2}\sin^2(\lambda\beta^{nc})V + \frac{p_\phi^2}{2V} \tag{30}$$

The non-commutative effective field equations are

$$\dot\beta = 4\pi G\gamma\frac{\partial H_{eff}}{\partial V} = -\frac{3}{2\gamma\lambda}\sin^2(\lambda\beta^{nc}) - 4\pi G\gamma\frac{p_\phi^2}{2V^2} \tag{31}$$

$$\dot V = -4\pi G\gamma\frac{\partial H_{eff}}{\partial \beta} = \frac{3}{\gamma\lambda}V\sin(\lambda\beta^{nc})\cos(\lambda\beta^{nc}) \tag{32}$$

$$\dot\phi = \frac{\partial H_{eff}}{\partial p_\phi} = -\frac{3a\theta}{4\pi G\gamma^2\lambda}\sin(\lambda\beta^{nc})\cos(\lambda\beta^{nc}) + \frac{p_\phi}{V} \tag{33}$$

$$\dot p_\phi = -\frac{\partial H_{eff}}{\partial\phi} = 0 \tag{34}$$

In the limit $\theta \to 0$ we recover the commutative field equations. Due to the field equation for ϕ, we note that now the matter density $\rho = \frac{\dot\phi^2}{2}$ is not given only by $\frac{p_\phi^2}{V^2}$, but by

$$\rho^{nc} = \frac{1}{2}\left(-\frac{3a\theta}{4\pi G\gamma^2\lambda}\sin(\lambda\beta^{nc})\cos(\lambda\beta^{nc}) + \frac{p_\phi}{V}\right)^2 \tag{35}$$

In order to construct the Friedmann equation we would need to obtain a relation for p_ϕ in terms of $\dot\phi$, but since the field equation for $\dot\phi$ is now more involved, such relation can not be obtained; and so the Friedmann equation can not be constructed.

The relational evolution of V in terms of ϕ is now given by

$$\left(\frac{dV}{d\phi}\right)^2 = \left(\frac{3V}{\gamma\lambda}\right)^2\sin^2(\lambda\beta^{nc})\cos^2(\lambda\beta^{nc})\left[\frac{3\theta}{8\pi G\gamma^2\lambda^2}\sin^2(\lambda\beta^{nc})\cos^2(\lambda\beta^{nc}) + \frac{p_\phi}{V}\right]^{-2} \tag{36}$$

When taking $\theta \to 0$ this relation reduces to (20).

4.1. Addition of potential term

In the case of a scalar field with a potential term $V(\phi)$ we have

$$H^{nc}_{eff} = -\frac{3}{8\pi G\gamma^2\lambda^2}\sin^2(\lambda\beta^{nc})V + \frac{p_\phi^2}{2V} + V(\phi^{nc})V \tag{37}$$

The non-commutative effective field equations are

$$\dot\beta = 4\pi G\gamma\frac{\partial H_{eff}}{\partial V} = -\frac{3}{2\gamma\lambda}\sin^2(\lambda\beta^{nc}) - 4\pi G\gamma\frac{p_\phi^2}{2V^2} + 4\pi G\gamma V(\phi^{nc}) + 4\pi G\gamma b\theta V\frac{\partial V(\phi^{nc})}{\partial V} \tag{38}$$

$$\dot V = -4\pi G\gamma\frac{\partial H_{eff}}{\partial \beta} = \frac{3}{\gamma\lambda}V\sin(\lambda\beta^{nc})\cos(\lambda\beta^{nc}) \tag{39}$$

$$\dot\phi = \frac{\partial H_{eff}}{\partial p_\phi} = -\frac{3a\theta}{4\pi G\gamma^2\lambda}V\sin(\lambda\beta^{nc})\cos(\lambda\beta^{nc}) + \frac{p_\phi}{V} \tag{40}$$

$$\dot p_\phi = -\frac{\partial H_{eff}}{\partial\phi} = -V\frac{\partial V(\phi^{nc})}{\partial\phi} \tag{41}$$

These field equations reduce to the commutative ones when taking $\theta \to 0$. The matter density $\frac{\dot{\phi}^2}{2} + V(\phi^{nc})$ is given by

$$\rho^{nc} = \frac{1}{2}\left(-\frac{3a\theta}{4\pi G\gamma^2\lambda}\sin(\lambda\beta^{nc})\cos(\lambda\beta^{nc}) + \frac{p_\phi}{V}\right)^2 + V(\phi^{nc}) \tag{42}$$

In the particular case in which $a = 0$ the whole noncommutativity is encoded in the shifted variable $\phi^{nc} = \phi + b\theta V$, in this case we have the Hamiltonian

$$H^{nc}_{eff} = -\frac{3}{8\pi G\gamma^2\lambda^2}\sin^2(\lambda\beta)V + \frac{p_\phi^2}{2V} + V(\phi^{nc})V \tag{43}$$

The non-commutative effective field equations are

$$\dot{\beta} = 4\pi G\gamma\frac{\partial H_{eff}}{\partial V} = -\frac{3}{2\gamma\lambda}\sin^2(\lambda\beta) - 4\pi G\gamma\frac{p_\phi^2}{2V^2} + 4\pi G\gamma V(\phi^{nc}) + 4\pi G\gamma b\theta V\frac{\partial V(\phi^{nc})}{\partial V} \tag{44}$$

$$\dot{V} = -4\pi G\gamma\frac{\partial H_{eff}}{\partial \beta} = \frac{3}{\gamma\lambda}V\sin(\lambda\beta)\cos(\lambda\beta) \tag{45}$$

$$\dot{\phi} = \frac{\partial H_{eff}}{\partial p_\phi} = \frac{p_\phi}{V} \tag{46}$$

$$\dot{p}_\phi = -\frac{\partial H_{eff}}{\partial \phi} = -V\frac{\partial V(\phi^{nc})}{\partial \phi} \tag{47}$$

These field equations reduce to the commutative ones when taking $\theta \to 0$. The matter density $\frac{\dot{\phi}^2}{2} + V(\phi^{nc})$ is given by

$$\rho^{nc} = \frac{p_\phi^2}{2V^2} + V(\phi^{nc}) \tag{48}$$

In this particular case, it is possible to construct the Friedmann equation, which takes the form

$$\mathrm{H}^2 = \left(\frac{\dot{V}}{3V}\right)^2 = \frac{8\pi G^2}{3}\rho^{nc}\left(1 - \frac{\rho^{nc}}{\rho_{max}}\right) \tag{49}$$

This equation reduces to the standard one by taking $\theta \to 0$.

The Klein-Gordon equation for the scalar field takes the form

$$\ddot{\phi} = -3H\dot{\phi} - \frac{1}{V}\frac{\partial V(\phi^{nc})}{\partial V} \tag{50}$$

which reduces to the standard one by taking $\theta \to 0$. The relational evolution of V in terms of ϕ is in this case the same as the commutative one since the field equations for V and ϕ remain unchanged.

4.2. First order non-commutative quantum corrections

In the following we will neglect terms of $\mathcal{O}(\theta^2)$.

Expanding the Hamiltonian (30) in λ, keeping only up to the leading term (λ^2), we have

$$H^{nc}_{eff} = -\frac{3V}{8\pi G\gamma^2}(\beta^2 + 2a\theta\beta p_\phi) + \frac{V\lambda^2}{8\pi G\gamma^2}(\beta^4 + 4a\theta\beta^3 p_\phi) + \frac{p_\phi^2}{2V} \tag{51}$$

This Hamiltonian incorporates the first quantum correction as well as the first non-commutative correction. The first addend is composed of the classical term β^2 plus its first non-commutative correction; whereas the second addend is composed of the first quantum correction (β^4 term) plus its first non-commutative correction.

The field equations are

$$\dot{\beta} = 4\pi G\gamma\frac{\partial H_{eff}}{\partial V} = -\frac{3}{2\gamma}(\beta^2 + 2a\theta\beta p_\phi) + \frac{\lambda^2}{2\gamma}(\beta^4 + 4a\theta\beta^3 p_\phi) - 4\pi G\gamma\frac{p_\phi^2}{2V^2} \tag{52}$$

$$\dot{V} = -4\pi G\gamma\frac{\partial H_{eff}}{\partial \beta} = \frac{3V}{\gamma}(\beta + a\theta p_\phi) - \frac{2V\lambda^2}{\gamma}(\beta^3 + 3a\theta\beta^2 p_\phi) \tag{53}$$

$$\dot{\phi} = \frac{\partial H_{eff}}{\partial p_\phi} = -\frac{3a\theta\beta V}{4\pi G\gamma^2} + \frac{a\lambda^2\theta\beta^3 V}{2\pi G\gamma^2} + \frac{p_\phi}{V} \tag{54}$$

$$\dot{p}_\phi = -\frac{\partial H_{eff}}{\partial \phi} = 0 \tag{55}$$

When taking $\theta \to 0$ these field equations reduce to the ones incorporating first order quantum corrections.

The energy density $\frac{\dot{\phi}^2}{2}$ is in this case given by

$$\rho = \frac{p_\phi^2}{2V^2} + a\theta\frac{\beta p_\phi}{\pi G\gamma^2}\left(\frac{\beta^2}{2} - \frac{3}{4}\right) \tag{56}$$

Employing the Hamiltonian constraint we have

$$\mathrm{H}^2 = \frac{1}{\gamma^2}\left(\beta^2 + 2a\theta\beta p_\phi\right) - \frac{4\lambda^2}{3\gamma^2}\left(\beta^4 + 4a\theta\beta^3 p_\phi\right) = \frac{8\pi G}{3}(\rho + \rho^\theta)\left(1 - \frac{\rho + \rho^\theta}{\rho_{max}}\right) \tag{57}$$

where $\rho^\theta = a\theta\left(\frac{3\beta p_\phi}{4\pi G\gamma^2} - \frac{\lambda^2\beta^3 p_\phi}{2\pi G\gamma^2}\right)$.

Equation (57) is the Friedmann equation with first order non-commutative and loop quantum corrections. We note that when $\theta \to 0$ we obtain the leading term of the modified Friedmann equation (19).

5. Discussion

A simple noncommutative extension of the open FLRW loop quantum cosmology has been constructed, by introducing a theta deformation at the effective scheme of Loop Quantum Cosmology. This model thus incorporates effective corrections from both Loop Quantum Cosmology and Noncommutative Geometry. In the general case it was not possible to construct the corresponding Friedmann equation; however it was possible to construct it for the first order non-commutative corrections. Also, a noncommutative modified Friedmann equation could be constructed for the case in which the noncommutativity is encoded just in the matter degree of freedom. The physical implications of such noncommutative corrections are currently being investigated and will be reported elsewhere.

In the general case (eq. (30)) it is seen from the equation for $\dot{V}$ that a bounce ocurs when $\beta = \frac{\pi}{2\pi} - a\theta p_\phi$, morover, at this value of β, the noncommutative density is equal to the value given by Effective Loop Quantum Comology, ρ_{max}. This means that the noncommutative extension constructed in this work does not modify the way in which a bounce is reached; however, the time at which a bounce ocurs can be shifted by tunning the noncommutative parameter θ. It is also seen from the equation for ρ^{nc}, that this new energy density incorporates a geometric (gravitational) term, which could kill the usual matter term for particular values in the noncommutative parameter. Naturally, the same conclusions can be drawn from the first order approximation (eq. (51)). For the particular case in which the whole noncommutativity is encoded in the matter degree of fredom (eq. (43)) we observe that noncommutativity amounts to only modifying the potential term, and therefore there are no modifications to the Effective Loop Quantum Cosmology scheme; we can conclude that in order to obtain a real modification of Effective Loop Quantum Comology, noncommutativity should be implemented in the gravitational degrees of freedom. However, it is interesting to investigate if a cosmological constant term could arise when implementing noncommutativity only in the matter sector.

Acknowledgments
This work was partially supported by CONACYT 167335, 179881 grants. PROMEP grants UGTO-CA-3. A.E.G. is supported by a CONACYT graduate fellowship.

[1] H.S. Snyder, *Phys. Rev.*,**71**,38, (1947). *Quantized space-time*
[2] F. Bayen, M. Flato, C. Fronsdal, A. Lichnerowicz and D. Sternheimer, *Lett. Math. Phys.* **1** (1977), 521530; F. Bayen, M. Flato, C. Fronsdal, A. Lichnerowicz and D. Sternheimer *Ann. Phys.* **111**, 61 (1978); F. Bayen, M. Flato, C. Fronsdal, A. Lichnerowicz and D. Sternheimer *Ann. Phys.* **111**, 111 (1978).
[3] J. Moyal, *Proc. Camb. Phil. Soc.* **45**, 99 (1949).
[4] Giuseppe Dito and Daniel Sternheimer, *Deformation quantization: genesis, developments and metamorphoses*, [arXiv:math/0201168].
[5] A. Connes, *Noncommutative Geometry*, Academic Press, New York (1994)
[6] A. Connes, M.R. Douglas and A. Schwarz, *J. High Energy Phys.* **02**, 003 (1998) *Noncommutative geometry and matrix theory*, [arXiv:hep-th/97111262].
[7] N. Seiberg and E. Witten, *J. High Energy Phys.* **9906**, 032 (1999) *String theory and noncommutative geometry*, [arXiv:hep-th/9908142].
[8] M. R. Douglas, N. A. Nekrasov, *Rev. Mod. Phys.* **73**, 977. *Noncommutative field theory*, [arXiv:hep-th/0106048]; R. J. Szabo, *Phys. Rept.* **378**, 207 (2003) *Quantum field theory on noncommutative spaces*, [arXiv:hep-th/0109162].
[9] B.S. DeWitt, *Gravitation: an Introduction to Current Research* (1962)
[10] J.W. Moffat, *Phys. Lett. B* **491** 345 (2000); A.H. Chamsedine, *Commun. Math. Phys.* **218**, 283 (2001); A.H. Chamsedine, *Commun. Math. Phys.* **218**, 283 (2001); H. Garcia-Compean, O. Obregon, C. Ramirez and M. Sabido, *Phys. Rev. D* **68**, 044015 (2003).
[11] H. Garcia-Compean, O. Obregon, and C. Ramirez, *Phys. Rev. Lett.*, **88**, 161301 (2002).
[12] L. Mezincescu, *Star operation in Quantum Mechanics*, [arXiv:hep-th/0007046]; J. Gamboa, M. Loewe and J.C. Rojas, *Phys. Rev. D* **64** 067901 (2001); M. Chaichian, M.M. Sheikh-Jabbari and A. Tureanu, *Phys. Rev. Lett.* **86**, 2716 (2001);
[13] G. D. Barbosa and N. Pinto-Neto, *Phys. Rev. D* **70**, 103512 (2004). *Noncommutative geometry and cosmology*, [arXiv:hep-th/0407111].
[14] W. Guzman, M. Sabido and J. Socorro, *Phys. Rev. D* **76**, 08732 (2007); J. Socorro, Luis O. Pimentel, C. Ortiz, M. Aguero, *Int. J. Theor. Phys.* **48**, 3567 (2009);
[15] J.C. López-Domínguez, O. Obregón, M. Sabido and C. Ramírez *Phys. Rev. D* **74**, 084024 (2006)
[16] A. Ashtekar and J. Lewandowski, Background independent quantum gravity: A status report, Class. Quant. Grav. **21**, R53-R152 (2004).
[17] T. Thiemann, Introduction to Modern Canonical Quantum General Relativity (Cambridge University Press, Cambridge, (2007)).
[18] M. Bojowald, Loop quantum cosmology, Living Rev. Rel. **8**, 11 (2005) [arXiv:gr-qc/0601085].
[19] A. Ashtekar and P. Singh, Loop quantum cosmology: A status report, Class. Quant. Grav. **28**, 213001 (2011);
[20] Ashtekar A, Pawlowski T and Singh P 2006 Quantum nature of the big bang: improved dynamics Phys. Rev. D **74** 084003
[21] A. Corichi and A. Karami, Loop quantum cosmology of k=1 FRW: A tale of two bounces, Phys. Rev. D **84**, 044003 (2011) [arXiv:1105.3724].
[22] K. Vandersloot, Loop quantum cosmology and the k=- 1 RW model, Phys. Rev. D **75**, 023523 (2007) [gr-qc/0612070].
[23] D. Chiou, Loop quantum cosmology in Bianchi Type I Models: Analytical Investigation, Phys. Rev D **75**, 024029 (2007); Effective dynamics, big bounces and scaling symmetry in Bianchi I loop quantum cosmology, Phys. Rev. D **76**, 124037 (2007).
[24] A. Ashtekar and E. Wilson-Ewing, Loop quantum cosmology of Bianchi type I models, Phys. Rev. D **79** 083535 (2009).
[25] A. Ashtekar and E. Wilson-Ewing, Loop quantum cosmology of Bianchi type II models, Phys. Rev. D80 123532 (2009).
[26] E. Wilson-Ewing, Loop quantum cosmology of Bianchi type IX models, Phys. Rev. D **82** 043508 (2010).
[27] M. Bojowald, Absence of singularity in loop quantum cosmology, Phys. Rev. Lett. **86**, 5227-5230 (2001).
[28] Garay L.J., Martín-Benito M., Mena Marugán, G.A., Inhomogeneous loop quantum cosmology: hybrid quantization of the Gowdy model, Phys. Rev. D **82** (2010), 044048. [arXiv:1005.5654].
[29] Ashtekar A and Schilling T A 1999 Geometrical formulation of quantum mechanics On Einsteins Path: Essays in Honor of Engelbert Schücking ed A Harvey (New York: Springer) pp 2365 [arXiv:gr-qc/9706069]
[30] Cosmas Zachos, *Int. J. Mod. Phys. A*, **17**, 297 (2002).
[31] H. Weyl, *The Theory of Groups and Quantum Mechanics* Dover, New York (1931).

[32] J. Gamboa, M. Loewe and J.C. Rojas, *Phys. Rev. D* **64** 067901 (2001); M. Chaichian, M.M. Sheikh-Jabbari and A. Tureanu, *Phys. Rev. Lett.* **86**, 2716 (2001).
[33] Sayipjamal Dulat and Kang Li *Commutator Anomaly in Noncommutative Quantum Mechanics*, [arxiv:hep-th/0508060].
[34] Aschieri, P. et al., *Noncommutative Sapcetimes: Symmetries in Noncommutative Geometry and Field Theory, Lect. Notes Phys.*, Springer (2009).
[35] P.M. Ho and H.C. Kao, *Phys. Rev. Lett.* **88**, 151602, (2002). *Phys. Rev. Lett.* **88**, 161301 (2002).

Journal of Physics: Conference Series **654** (2015) 012004 doi:10.1088/1742-6596/654/1/012004

Relativistic bulk viscosity in the relaxation time approximation: a chaotic velocities approach

A. L. García-Perciante[1], A. R. Méndez[2], A. Sandoval-Villalbazo[3]

[1, 2]Depto. de Matemáticas Aplicadas y Sistemas, Universidad Autónoma Metropolitana-Cuajimalpa, Prol. Vasco de Quiroga 4871, México D.F 05348, México.
[3]Depto. de Física y Matemáticas, Universidad Iberoamericana, Prolongación Paseo de la Reforma 880, México D. F. 01219, México.

E-mail: [1] `algarcia@correo.cua.uam.mx`

E-mail: [2] `amendez@correo.cua.uam.mx`

E-mail: [3] `alfredo.sandoval@ibero.mx`

Abstract. In this short note, the bulk viscosity for a high temperature dilute gas is calculated by applying the Chapman-Enskog method within Marle's relaxation time approximation. The expression for the stress-tensor established in Ref.[1], using explicitly the concept of chaotic velocity, is used in order to obtain the transport coefficient. The result is compared with previous expressions obtained by other authors using similar methods and emphasis is made on the agreement when a corrected relaxation parameter is considered.

1. Introduction

Transport coefficients are of current interest in high temperature gases, when the thermal energy becomes comparable or exceeds the rest energy of the individual particles. Other authors have calculated these gas parameters using various methods within kinetic theory. In this work we introduce the simplest model for the collisional term of the relativistic Boltzmann's equation, namely Marle's model, in the particle's frame and follow a direct method analogous to the one employed in the BGK approximation. Following such procedure, one can obtain directly the part of the distribution function which is coupled to the derivatives of the hydrodynamic velocity, making possible the calculation of the corresponding bulk viscosity. The expression for the stress tensor here considered corresponds to the one established in Ref.[1] by means of explicitly introducing the concept of chaotic velocity in relativistic kinetic theory. In such formalism, Lorentz transformations are used in order to write thermodynamic fluxes in an arbitrary reference frame in terms of integrals which can be calculated in a comoving one.

An analytic expression for the coefficient, which vanishes in the non-relativistic low temperature limit, is established following the steps described above. The result is compared with values obtained by Cercignani & Kremer [2] using the complete collision kernel and a procedure akin to a Maxwellian iteration, within Marle's approximation in the method of moments. The results obtained in the present work are consistent with those presented in Ref.[2] and a better adjustment in the ultra-relativistic case is archived when the adjusted relaxation time established in Ref.[3] is considered. The coefficient here obtained, together with previous ones for other fluxes may be useful for addressing hot dilute gases in dissipative regimes.

In order to accomplish the task described above we have divided the rest of the work in three sections. In Section 2 we state the relativistic Boltzmann equation and outline the steps leading to the deviation from the equilibrium distribution function within the Chapman-Enskog hypothesis. Section 3 is devoted to the explicit calculation of the bulk viscosity coefficient. A brief discussion and final remarks are included in Section 4.

2. The relativistic Boltzmann equation within Marle's approximation

The starting point of the calculation is the relativistic Boltzmann equation with a relaxation time approximation:

$$v^\alpha f_{,\alpha} = -\frac{f - f^{(0)}}{\tau} \tag{1}$$

where v^α is the molecular velocity measured in an arbitrary reference frame and τ is a parameter related to characteristic time (see Ref. [3]) in which collisions relax the distribution function f to its equilibrium value $f^{(0)}$. For the relativistic dilute gas, $f^{(0)}$ is given by

$$f^{(0)} = \frac{n}{4\pi c^3 z K_2\left(\frac{1}{z}\right)} e^{\frac{\mathcal{U}_\beta v^\beta}{zc^2}} \tag{2}$$

where n is the particle number density and $\mathcal{U}_\beta$ the hydrodynamic velocity. An account of the basic properties of the Jüttner distribution function can be found in Refs.[2, 4, 5, 6]. These, together with the temperature T, which corresponds to the local tempereature, measured in the comoving frame of any volume element of the fluid and is included in the relativistic parameter $z = k_B T/mc^2$, are the state variables describing the gas. Also, c is the speed of light, k_B is Boltzmann's constant and K_n the $n - th$ modified Bessel function of the second kind .

Introducing the Chapman-Enskog hypothesis, $f = f^{(0)} + f^{(1)}$ in Eq. (1) and following the standard procedure (see for example Ref.[7]) one obtains, for the non-equilibrium part of the distribution function

$$f^{(1)} = -\tau f^{(0)} \left\{ v^\beta h^\alpha_\beta \left[\frac{n_{,\alpha}}{n} + \frac{T_{,\alpha}}{T} \left(1 - \frac{v^\mu \mathcal{U}_\mu}{zc^2} - \frac{\mathcal{G}\left(\frac{1}{z}\right)}{z} \right) + \frac{v_\mu}{zc^2} \mathcal{U}^\mu_{;\alpha} \right] \right. \\ \left. + \left(\frac{v^\beta \mathcal{U}_\beta}{c^2} \right) \left[\mathcal{U}^\mu_{;\mu} + \frac{k_B}{C_n} \mathcal{U}^\nu_{;\nu} \left(1 - \frac{v^\mu \mathcal{U}_\mu}{zc^2} - \frac{\mathcal{G}\left(\frac{1}{z}\right)}{z} \right) + \frac{p_{,\nu} h^{\mu\nu} v_\mu}{mn\mathcal{G}\left(\frac{1}{z}\right) zc^2} \right] \right\} \tag{3}$$

where $\mathcal{G}\left(\frac{1}{z}\right) = K_3\left(\frac{1}{z}\right)/K_2\left(\frac{1}{z}\right)$, $p = nkT$ is the hydrostatic pressure and C_n is the heat capacity. In order to obtain Eq. (3) use has been made of the following decomposition for the molecular velocity

$$v^\alpha = v^\beta h^\alpha_\beta - \left(\frac{v^\beta \mathcal{U}_\beta}{c^2} \right) \mathcal{U}^\alpha \tag{4}$$

where $h^{\mu\nu} = \eta^{\mu\nu} + \mathcal{U}^\mu \mathcal{U}^\nu / c^2$ is the so-called spatial projector. Also, the Euler equations have been used in order to write proper time derivatives in terms of spatial gradients.

Equation (4) constitutes an explicit expression for the first order in the gradients correction to the equilibrium distribution function. This attractive feature, the fact that the distribution function can be solved for analytically, makes this procedure particularly simple and useful in order to establish the structure of constitutive equations in a somewhat straightforward manner and the basic dependence of transport coefficients with state variables.

In order to address viscous dissipation, we start by noticing that the vector terms, containing $n_{,\alpha}$ $T_{,\alpha}$ and $p_{,\alpha}$ will not contribute to the stress tensor. This can be easily seen by examining the parity in the integral leading to the momentum flux average. Also, Curie's principle states that only forces and fluxes of the same tensor rank are coupled in constitutive equations. Thus, the part of the distribution function that will contribute to viscous dissipation is given by

$$-\tau f^{(0)} \left\{ h^\alpha_\beta \frac{v^\beta v_\mu}{zc^2} \mathcal{U}^\mu_{;\alpha} + \left(\frac{v^\mu \mathcal{U}_\mu}{c^2} \right) \left[1 + \frac{k_B}{C_n} \left(1 - \frac{v^\beta \mathcal{U}_\beta}{zc^2} - \frac{\mathcal{G}\left(\frac{1}{z}\right)}{z} \right) \right] \mathcal{U}^\nu_{;\nu} \right\}. \tag{5}$$

In particular, writing $\mathcal{U}^\mu_{;\alpha}$ as the sum of a symmetric traceless part, an antisymmetric part and a scalar part, the contribution arising from the first term in Eq. (5) to the trace of the stress tensor is given by

$$\frac{1}{3} \frac{1}{zc^2} h^\alpha_\beta \eta^\mu_\alpha v_\mu v^\beta \mathcal{U}^\ell_{;\ell} \tag{6}$$

where $\ell = 1, 2, 3$, and thus, the part of the distribution function proportional to the divergence of the hydrodynamic velocity is given by

$$f^{(1)}_b = -\tau f^{(0)} \left\{ \frac{1}{3zc^2} h^\alpha_\beta \eta^\mu_\alpha v_\mu v^\beta + \left(\frac{v^\beta \mathcal{U}_\beta}{c^2} \right) \left[1 + \frac{k_B}{C_n} \left(1 - \frac{v^\beta \mathcal{U}_\beta}{zc^2} - \frac{\mathcal{G}\left(\frac{1}{z}\right)}{z} \right) \right] \right\} \mathcal{U}^\ell_{;\ell} \tag{7}$$

where $\ell = 1, 2, 3$. As extensively discussed in Refs. [1, 8], the invariants included in the expression above can be calculated in a local rest frame moving with the fluid element where the molecules' velocities are given by their chaotic, or peculiar, component which we denote by $K^\mu = \gamma_k \left[\vec{k}, c\right]$ such that

$$\frac{1}{3}\frac{1}{zc^2} h^\alpha_\beta \eta^\mu_\alpha v_\mu v^\beta = \frac{k^2 \gamma_k^2}{3zc^2} \tag{8}$$

and

$$v^\mu \mathcal{U}_\mu = -\gamma_k c^2. \tag{9}$$

Chaotic velocity refers to molecular velocity measured by an observer comoving with the local hydrodynamic velocity. This variable allows for the calculation of invariant integrals in phase space and for a clear definition of state variables and dissipative fluxes in a representation analogous to the usual one in the non-relativistic case [9]. Introducing these expressions in Eq. (7) one obtains

$$f_b^{(1)} = -\tau f^{(0)} \left\{ \frac{k^2\gamma_k^2}{3zc^2} - \gamma_k \left[1 + \frac{k_B}{C_n} \left(1 - \frac{v^\beta \mathcal{U}_\beta}{zc^2} - \frac{\mathcal{G}\left(\frac{1}{z}\right)}{z} \right) \right] \right\} \mathcal{U}^\ell_{;\ell} \tag{10}$$

which will be used in the following section in order to obtain the bulk viscosity within this framework. Notice that, since

$$\frac{k_B}{C_n} = \frac{z^2}{\left[1 + 5z\mathcal{G}\left(\frac{1}{z}\right) - \mathcal{G}^2\left(\frac{1}{z}\right) - z^2\right]} \tag{11}$$

$f_b^{(1)}$ vanishes in the non-relativistic limit, as is also the case in the kinetic theory treatment starting with the non-relativistic Boltzmann equation and the Maxwell-Boltzmann equilibrium distribution.

3. Relativistic bulk viscosity in Marle's approximation

As mentioned in the introduction, the expression for the stress tensor here used in order to calculate the relativistic bulk viscosity is the one established in Ref.[1] namely,

$$\Pi^{\mu\nu} = m h^\mu_\alpha h^\nu_\beta \mathcal{L}^\alpha_\eta \mathcal{L}^\beta_\delta \int f^{(1)} K^\eta K^\delta d^*K. \tag{12}$$

Here $\mathcal{L}^{\mu\nu}$ is a Lorentz boost with velocity $\vec{u}$, that is a transformation from an arbitrary reference frame to the frame where the hydrodynamic velocity vanishes. Following the usual procedure, the stress tensor can be written as

$$\Pi^{\mu\nu} = (\Pi^{\mu\nu})^{(a)} + (\Pi^{\mu\nu})^{(s)} + \frac{1}{3}\Pi^{\alpha\beta}\eta_{\alpha\beta}\eta^{\mu\nu} \tag{13}$$

where the first term corresponds to the antisymmetric part, the second to the symmetric part and the third to the trace. It is this third term that will be coupled to the divergence of the hydrodynamic velocity. The proposed constitutive equation is thus

$$\frac{1}{3}\eta_{\nu\mu}\Pi^{\nu\mu} = -\eta \mathcal{U}^\mu_{;\mu} \tag{14}$$

where η is the bulk viscosity. Thus,

$$\eta = \frac{m}{3}\tau \eta_{\nu\mu} \mathcal{O}^\mu_a \mathcal{O}^\nu_b \int f^{(0)} \left\{ \frac{k^2\gamma_k^2}{3zc^2} - \gamma_k \left[1 + \frac{k_B}{C_n}\left(1 + \frac{\gamma_k}{z} - \frac{\mathcal{G}\left(\frac{1}{z}\right)}{z} \right) \right] \right\} K^a K^b d^*K \tag{15}$$

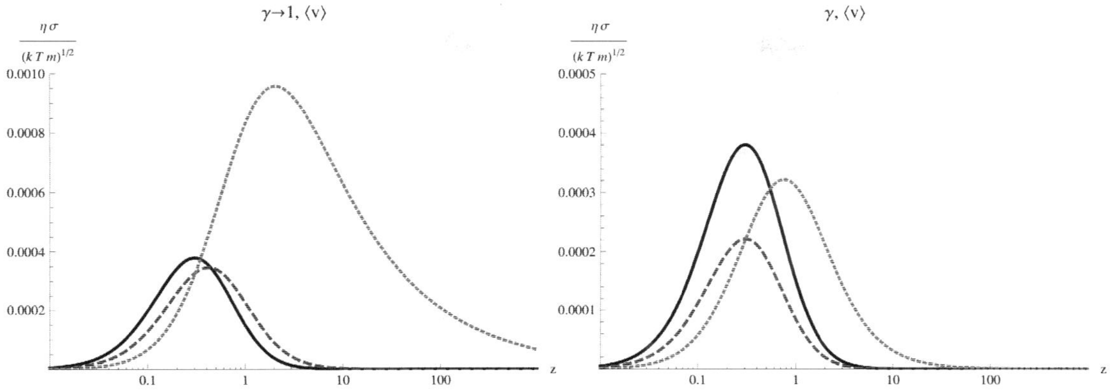

Figure 1. Normalized bulk viscosity with $\mathcal{V} = \langle k \rangle$(left) and $\mathcal{V} = \langle \gamma_k \rangle \langle k \rangle$ (right). The dotted line shows the result obtained in the present paper while the dashed curve corresponds to value obtained using the method of moments in Ref.[2]. The continuous line corresponds to the value obtained by using the complete kernel in the hard sphere's case [2]. Should be pointed out that in the non-relativistic limit $\langle \gamma_k \rangle \sim 1 + z$.

where the tensor $\mathcal{O}^{\mu\nu} = h^{\mu}_{\alpha}\mathcal{L}^{\alpha\nu}$ has the property $\mathcal{O}^{\mu 4} = 0$. Other properties of this tensor can be found in Ref. [1]. Also, since all $a \neq b$ lead to vanishing integrals and changing the variable to γ_k one obtains

$$\eta = 4\pi c^3 \frac{m}{9}\tau \int f^{(0)} \left\{ \frac{k^2\gamma_k^2}{3zc^2} - \gamma_k \left[1 + \frac{k_B}{C_n} \left(1 + \frac{\gamma_k}{z} - \frac{\mathcal{G}\left(\frac{1}{z}\right)}{z} \right) \right] \right\} \gamma_k^2 k^2 \sqrt{\gamma_k^2 - 1} d\gamma_k. \tag{16}$$

The corresponding integrals can be readily calculated and simplifying one obtains

$$\eta = \frac{1}{3} nmc^2\tau \frac{\left[z\left(2 - 20z^2\right)\mathcal{G}\left(\frac{1}{z}\right) + 13z^2\mathcal{G}^2\left(\frac{1}{z}\right) - 2z\mathcal{G}^3\left(\frac{1}{z}\right) - 3z^2 \right]}{3\left[1 + 5z\mathcal{G}\left(\frac{1}{z}\right) - \mathcal{G}^2\left(\frac{1}{z}\right) - z^2 \right]} \tag{17}$$

which in the low z, non-relativistic limit, vanishes:

$$\eta \approx \frac{5}{18} nmc^2\tau z^3. \tag{18}$$

In Figs. (1) and (2) the results here obtained are shown together with those obtained by Cercignani and Kremer in Ref.[2] using the complete collision kernel with hard spheres' model and Marle's approximation with the method of moments. The relaxation time parameter used for the plots is calculated as:

$$\tau = \frac{1}{4n\pi\sigma\mathcal{V}} \tag{19}$$

where $\mathcal{V}$ is a characteristic speed given by either the sound speed v_s, a mean velocity $\langle k \rangle$ or either of those values multiplied by an adjusting factor $\langle \gamma_k \rangle$ corresponding to Marle's approximation, as broadly discussed in Ref.[3]. The choice of these velocities, as pointed out in Ref. [2], arises from the fact that both v_s and $\langle k \rangle$ are of the same order as the average of the relative velocity, which is used to calculate τ in the non-relativistic case.

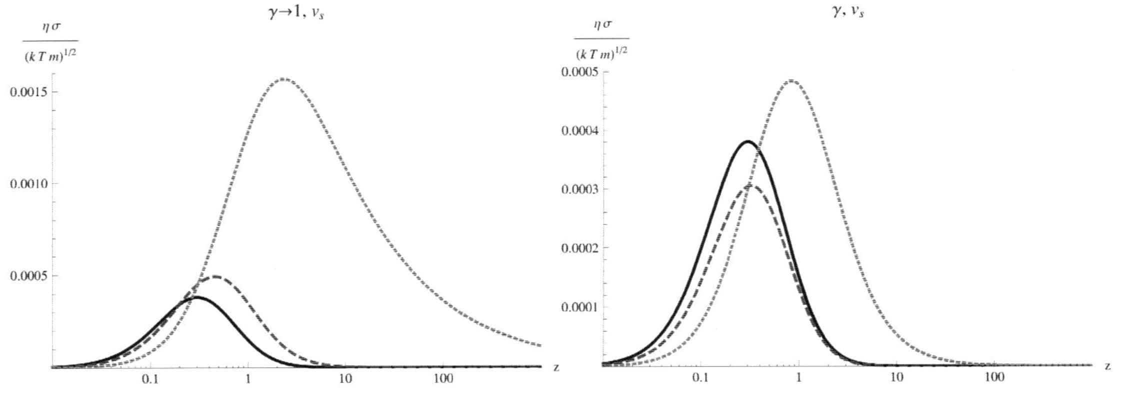

Figure 2. Normalized bulk viscosity with $\mathcal{V} = v_s$(left) and $\mathcal{V} = \langle\gamma_k\rangle\, v_s$ (right). The dotted line shows the result obtained in the present paper while the dashed curve corresponds to value obtained using the method of moments in Ref.[2]. The continuous line corresponds to the value obtained by using the complete kernel in the hard sphere's case [2].

4. Final Remarks

Viscous dissipation in non-relativistic simple fluids is only due to shear stresses. As temperature increases relativistic effects in molecular dynamics lead to a non-vanishing bulk viscosity which increases dissipation. This effect has important consequences on cosmological fluid dynamics since bulk viscosity is a source of entropy production in homogeneous and isotropic space-times [10].

In this work we obtained the bulk viscosity coefficient as a function of the relativistic parameter z using a relaxation time approximation, namely Marle's model. A similar calculation is carried out by Cercignani & Kremer in Ref.[2] using the method of moments. The novel feature of this work is the use of the standard Chapman-Enskog procedure and the concept of chaotic velocity, included both in the deviation from equilibrium of the distribution function and the expression for the stress-energy tensor in terms of Lorentz transformations. It should be emphasized that the relaxation equation is here used directly without any additional iterative procedures or corrections. The result obtained features good agreement when the correcting factor introduced in Ref.[3] for the characteristic time is considered. Moreover, the consideration of the $\langle\gamma_k\rangle$ factor in the expression for τ improves the behavior of the transport coefficient obtained here within Marle's model as well as the one calculated in Ref.[2].

The behavior of the bulk viscosity is shown in Figs. (1) and (2). The coefficient vanishes in the low temperature limit, grows with temperature and decreases after reaching a peak value for thermal energies comparable to the particles' rest mass energy. In the ultra-relativistic limit, the coefficient asymptotically approaches zero. The physics behind this behavior is an interesting subject and will be addressed elsewhere.

Acknowledgments

The authors acknowledge support from CONACyT through grant number CB2011/167563.

References

[1] García-Perciante A. L., Sandoval-Villalbazo A., García-Colín L. S, Jour. Non-Equilib. Thermodyn. **37**, 43 (2012).

[2] C. Cercignani and G. Medeiros Kremer, *The relativistic Boltzmann equation: theory and applications* (BirkhauserVerlag, Basel, 2002).

[3] A. R. Méndez and A. L. García-Perciante, Relativistic heat conduction: the kinetic theory approach and comparison with Marle's model, AIP Conf. Proc. 1312, 65 (2010).

[4] de Groot, S.R., Van Leeuwen, W.A., Van der Wert, Ch.: Relativistic Kinetic Theory. North Holland Publ. Co., Amsterdam (1980)

[5] L. S. García-Colín, G. Chacón-Acosta, L. Dagdug-Lima (Editors): *100 años de teoría cinética relativista*, Universidad Autónoma Metropolitana (2013).

[6] David Cubero, Jesús Casado-Pascual, Jörn Dunkel, Peter Talkner, and Peter Hänggi Phys. Rev. Lett. 99, 170601.

[7] S. Chapman, and T. G. Cowling, *The mathematical theory of non-uniform gases*, Cambridge Mathematical Library, United Kingdom, 1971 (3rd. ed).

[8] V. Moratto, A. L. García-Perciante, AIP Conference Proceedings 1578, 86 (2014).

[9] A.L. García-Perciante, AIP Conf. Proc. 1332, 216 (2011).

[10] J. Bernstein, *Kinetic Theory in the Expanding Universe*, Cambridge Univestiy Press, 1988.

Lagrangian formulation for fermionic and supersymmetric non conservative systems

NE Martínez-Pérez and C Ramírez
Benemérita Universidad Autónoma de Puebla, Facultad de Ciencias Físico Matemáticas, P.O. Box 165, 72000 Puebla, México.
E-mail: nephtalieliceo@hotmail.com, cramirez@fcfm.buap.mx

Abstract. We show that a recent formulation of the principle of stationary action, compatible with generic non conservative interactions, is suitable for fermionic and supersymmetric systems. The main features of the mechanics contained in this new action principle are also encountered in the examples discussed in this work.

1. Introduction

The evolution of mechanical systems can be irreversible and non invariant under time reversal. Such systems are studied usually by means of their equations of motion. In two recent papers, Galley [1, 2] gives a systematic proposal for a lagrangian formulation for nonconservative systems. This proposal amounts to a modification of Hamilton's variational principle, introducing a nonconservative "potential". This formalism is inspired on the closed time path formalism, originally proposed by Schwinger [3] for field theory. Supersymmetry is a symmetry which has allowed consistent formulations of unification theories of all fundamental forces. It imposes rather strong constraints on the particle content, mass spectrum, and interactions. Thus we consider interesting to explore if supersymetry is compatible with this formulation for nonconservative systems [4]. The generalization for supersymmetric systems is done in the superspace formalism. The boundary conditions must be also modified, and are given in terms of superfields. We illustrate the formalism by some examples.

2. Galley's mechanics for non conservative systems

Galleys formulation is based on the doubling of the degrees of freedom of a conservative system with Lagrangian $L(q, \dot{q})$, $q \to (q_1, q_2)$. The action is defined as the integral of a conservative Lagrangian over two trajectories in configuration space, each one for one of the two copies of the original degrees of freedom. To this action, it is added the integral of a generalized potential $K(q_1, \dot{q}_1, q_2, \dot{q}_2)$, that couples these two trajectories. This potential is antisymmetric under the interchange $q_1 \leftrightarrow q_2$. Thus there is a new Lagrangian

$$\Lambda = L(q_1, \dot{q}_1, t) - L(q_2, \dot{q}_2, t) + K(q_1, \dot{q}_1, q_2, \dot{q}_2). \tag{1}$$

The variation of this action is given under the boundary conditions at the initial and final times

$$\delta q_1(t_i) = \delta q_2(t_i) = 0, \tag{2}$$

$$\delta q_1(t_f) = \delta q_2(t_f), \qquad q_1(t_f) = q_2(t_f), \qquad \dot{q}_1(t_f) = \dot{q}_2(t_f). \tag{3}$$

The last three equations define the so called *equality condition*. Thus, the equations of motion are

$$\frac{\partial}{\partial q_1}(L+K) - \frac{d}{dt}\frac{\partial}{\partial \dot{q}_1}(L+K) = 0, \tag{4}$$

$$\frac{\partial}{\partial q_2}(L-K) - \frac{d}{dt}\frac{\partial}{\partial \dot{q}_2}(L-K) = 0. \tag{5}$$

Taking into account the antisymmetry of the potential, and making the *physical limit* $q_1 = q_2 = q$, (17) and (14) coincide, giving the nonconservative equations of motion

$$\frac{\partial L}{\partial q} - \frac{d}{dt}\frac{\partial L}{\partial \dot{q}} = F_K \equiv \left(\frac{\partial}{\partial q_2} - \frac{d}{dt}\frac{\partial}{\partial \dot{q}_2}\right) K(q_1, \dot{q}_1, q_2, \dot{q}_2)\Big|_{q_1=q_2=q}. \tag{6}$$

A convenient change of variables is given by

$$q_+ = \tfrac{1}{2}(q_1 + q_2), \tag{7}$$

$$q_- = q_1 - q_2. \tag{8}$$

In the physical limit, the variable q_- turns to zero and the variable q_+ reduces to the physical coordinate q. In addition, the equality condition gets a more practical statement in terms of this coordinates.

2.1. Dissipative systems

The doubling of the degrees of freedom allows us to introduce terms of the form $q_- \dot{q}_+^n$, which allow to construct "generalized potentials" $K(q_\pm, \dot{q}_\pm)$ for dissipative forces. The form of the lagrangian for a general system is

$$\Lambda = L(q_1) - L(q_2) + K(q_\pm). \tag{9}$$

A simple, but illustrative example is the damped harmonic oscillator, with the nonconservative lagrangian

$$\Lambda = \dot{q}_+\dot{q}_- - \omega^2 q_+ q_- + \frac{\epsilon}{2}(\dot{q}_- q_+ - \dot{q}_+ q_-), \tag{10}$$

which, after the variation as previously shown, leads to the well-known equation of motion

$$\ddot{q} + \epsilon\dot{q} + \omega^2 q = 0. \tag{11}$$

3. Fermionic degrees of freedom

Now we state formally Galley's action principle for a fermionic system. We define the action S as follows

$$S[\psi_a^\alpha] = \int_{t_i}^{t_f} \Lambda(\psi_i^\alpha, \dot{\psi}_i^\alpha), \tag{12}$$

where the generalized fermionic lagrangian is, in analogy with the bosonic case, is given in terms of the doubled variables ψ_1 and ψ_2

$$\Lambda[\psi_1^\alpha, \psi_2^\alpha] = L(\psi_1^\alpha, \dot{\psi}_1^\alpha) - L(\psi_2^\alpha, \dot{\psi}_2^\alpha) + K(\psi_a^\alpha, \dot{\psi}_a^\alpha). \tag{13}$$

The boundary conditions are the same as in the previous section

$$\delta\psi_1(t_i) = \delta\psi_2(t_i) = 0, \tag{14}$$

$$\psi_1(t_f) = \psi_2(t_f), \qquad \dot{\psi}_1(t_f) = \dot{\psi}_2(t_f), \qquad \delta\psi_1(t_f) = \delta\psi_2(t_f). \tag{15}$$

As pointed out by Teitelboim [5], the boundary conditions mus take into account that the equations of motion of fermions are first order, hence only one boundary condition is required. Thus, if ψ_1 and ψ_2 coincide at the final time, they must coincide also at the initial time, which should be added to (14) for consistency. Performing the variation of the action, the usual Euler-Lagrange equations are obtained

$$\frac{d}{dt}\left(\frac{\partial\Lambda}{\partial\dot{\psi}_a^\alpha}\right) - \frac{\partial\Lambda}{\partial\psi_a^\alpha} = 0. \tag{16}$$

3.1. Example: damped fermionic oscillator
Consider the lagrangian, given in terms of the $\pm$ variables:

$$\Lambda = z^*\left(i\dot{\bar{\psi}}_+\psi_- + i\dot{\psi}_-\bar{\psi}_+\right) + z\left(i\dot{\bar{\psi}}_-\psi_+ + i\dot{\psi}_+\bar{\psi}_-\right) + c\left(\bar{\psi}_-\psi_+ + \bar{\psi}_+\psi_-\right), \tag{17}$$

where $z = x + iy$ is a complex number. The equations of motion (taking the physical limit) are

$$2iz\dot{\psi} - \omega\psi = 0, \tag{18}$$
$$2iz^*\dot{\bar{\psi}} + \omega\bar{\psi} = 0, \tag{19}$$

whose solutions are given by the complex exponentials

$$\psi(t) = \psi_0 \exp\left[\frac{\omega(-ix-y)}{2(x^2+y^2)}t\right], \tag{20}$$
$$\bar{\psi}(t) = \bar{\psi}_0 \exp\left[\frac{\omega(ix-y)}{2(x^2+y^2)}t\right], \tag{21}$$

which correspond to a harmonic motion, with an exponentially decaying amplitude.

In terms of the coordinates $q_{1,2}$ we have

$$\Lambda = x\left(i\dot{\bar{\psi}}_1\psi_1 + i\dot{\psi}_1\bar{\psi}_1\right) + \omega\bar{\psi}_1\psi_1 - x\left(i\dot{\bar{\psi}}_2\psi_2 + i\dot{\psi}_2\bar{\psi}_2\right) - \omega\bar{\psi}_2\psi_2 + y\left[\dot{\bar{\psi}}_2\psi_1 - \dot{\bar{\psi}}_1\psi_2 - \dot{\psi}_2\bar{\psi}_1 + \dot{\psi}_1\bar{\psi}_2\right], \tag{22}$$

where we can see that this lagrangian fits Galley's prescription by means of the identifications

$$L = x\left(i\dot{\bar{\psi}}\psi + i\dot{\psi}\bar{\psi}\right) + \omega\bar{\psi}\psi, \tag{23}$$
$$K = y\left[\dot{\bar{\psi}}_2\psi_1 - \dot{\bar{\psi}}_1\psi_2 - \dot{\psi}_2\bar{\psi}_1 + \dot{\psi}_1\bar{\psi}_2\right]. \tag{24}$$

As required, this generalized (fermionic-) potential is antisymmetric under $\psi_1 \leftrightarrow \psi_2$.

4. Supersymmetric action

The supersymmetric version of Galley's action principle is presented in the $N = 2$ superspace formalism (some general remarks on superspace and notation conventions can be found in Apendix A). For simplicity, let us consider a system described by a single (superfield) supercoordinate $\Phi(t, \theta, \bar{\theta})$.

The non conservative supersymmetric lagrangian density is defined as

$$\Gamma(\Phi_1, \Phi_2) = \mathcal{L}(\Phi_1) - \mathcal{L}(\Phi_2) + \mathcal{K}(\Phi_1, \Phi_2). \tag{25}$$

where $\mathcal{K}$ is the non conservative "superpotential". The action is defined as

$$S = \int dX\ \Gamma(\Phi_1, \Phi_2). \tag{26}$$

In order to perform the variation of the action, the following boundary conditions are imposed

$$\delta\Phi_1|_{t_i} = \delta\Phi_2|_{t_i} = 0, \tag{27}$$

$$\Phi_1|_{t_f} = \Phi_2|_{t_f}, \qquad D\Phi_1|_{t_f} = D\Phi_2|_{t_f}, \qquad \bar{D}\Phi_1|_{t_f} = \bar{D}\Phi_2|_{t_f}, \qquad \delta\Phi_1|_{t_f} = \delta\Phi_2|_{t_f}. \tag{28}$$

According to the discussion in the fermionic case, we should impose the aditional condition $\Phi_1(X,\epsilon)|_{t_i} = \Phi_2(X,\epsilon)|_{t_i}$.

The variation of the action leads to the Euler-Lagrange equations

$$\frac{\partial\Gamma}{\partial\Phi_a} - D\frac{\partial\Gamma}{\partial D\Phi_a} - \bar{D}\frac{\partial\Gamma}{\partial\bar{D}\Phi_a} = 0. \tag{29}$$

In fact, the boundary terms of the previous variation are given by

$$\begin{aligned}
B.T. &= \int dX\ \left\{D\left(\eta_1\frac{\partial\Gamma}{\partial D\Phi_1} - \eta_2\frac{\partial\Gamma}{\partial D\Phi_2}\right) + \bar{D}\left(\eta_1\frac{\partial\Gamma}{\partial\bar{D}\Phi_1} - \eta_2\frac{\partial\Gamma}{\partial\bar{D}\Phi_2}\right)\right\} \\
&= i\neg D\left(\eta_1\frac{\partial\Gamma}{\partial D\Phi_1} - \eta_2\frac{\partial\Gamma}{\partial D\Phi_2}\right)\Bigg|_{t_i}^{t_f} - i\neg\bar{D}\left(\eta_1\frac{\partial\Gamma}{\partial\bar{D}\Phi_1} - \eta_2\frac{\partial\Gamma}{\partial\bar{D}\Phi_2}\right)\Bigg|_{t_i}^{t_f} \\
&= i\neg\left\{(D\eta_1)\frac{\partial(\mathcal{L}+\mathcal{K})}{\partial D\Phi_1} + \eta_1 D\frac{\partial(\mathcal{L}+\mathcal{K})}{\partial D\Phi_1} - (D\eta_2)\frac{\partial(\mathcal{L}-\mathcal{K})}{\partial D\Phi_2} - \eta_2 D\frac{\partial(\mathcal{L}-\mathcal{K})}{\partial D\Phi_2}\right\}\Bigg|_{t_i}^{t_f} \\
&- i\neg\left\{(\bar{D}\eta_1)\frac{\partial(\mathcal{L}+\mathcal{K})}{\partial\bar{D}\Phi_1} + \eta_1\bar{D}\frac{\partial(\mathcal{L}+\mathcal{K})}{\partial\bar{D}\Phi_1} - (\bar{D}\eta_2)\frac{\partial(\mathcal{L}-\mathcal{K})}{\partial\bar{D}\Phi_2} - \eta_2\bar{D}\frac{\partial(\mathcal{L}-\mathcal{K})}{\partial\bar{D}\Phi_2}\right\}\Bigg|_{t_i}^{t_f}.
\end{aligned} \tag{30}$$

In these expressions, $\neg$ means setting $\theta, \bar{\theta} = 0$ in the expressions to the right. Clearly, $\neg(\Phi\Psi) = (\neg\Phi)(\neg\Psi)$. With the last expression we can see that, because of the boundary conditions, all of these terms cancel or vanish.

4.1. Example: Damped supersymmetric oscillator

The lagrangian density in the $\pm$ variables is given by

$$\Gamma = \sqrt{m}\left[z^*\bar{D}\Phi_- D\Phi_+ + z\bar{D}\Phi_+ D\Phi_- - \omega\Phi_+\Phi_-\right], \tag{31}$$

where $z = \frac{1}{2}\left[\sqrt{1-(\frac{\epsilon}{2\omega})^2} + i\frac{\epsilon}{2\omega}\right]$.

The equation of motion, in the superspace language, is

$$[zD\bar{D} - z^*\bar{D}D]\Phi - \omega\Phi = 0. \tag{32}$$

In components, (32) is equivalent to the equations of motion of the damped (bosonic) oscillator (11), and those of the fermionic oscilators (18), (19). The solutions are

$$\psi(t) = e^{\left(-i\omega' - \frac{\epsilon}{2}\right)t}\psi_0, \tag{33}$$

$$\bar{\psi}(t) = e^{\left(i\omega' - \frac{\epsilon}{2}\right)t}\bar{\psi}_0, \tag{34}$$

$$q(t) = \left[e^{i\omega' t}q_{01} + e^{-i\omega' t}q_{02}\right]e^{-\frac{\epsilon t}{2}}, \tag{35}$$

where $\omega' = \omega\sqrt{1 - (\frac{\epsilon}{2\omega})^2}$.

In terms of the original coordinates we have

$$\Gamma = \frac{\sqrt{m}}{2\omega}\left[w'\bar{D}\Phi_1 D\Phi_1 - \omega^2\Phi^2 - \left(w'\bar{D}\Phi_2 D\Phi_2 - \omega^2\Phi_2^2\right) + i\frac{\epsilon}{2}\left(\bar{D}\Phi_2 D\Phi_1 - \bar{D}\Phi_1 D\Phi_2\right)\right], \quad (36)$$

so that, we can identify the supersymmetric potential

$$\mathcal{K} = i\frac{\epsilon\sqrt{m}}{4\omega}\left(\bar{D}\Phi_2 D\Phi_1 - \bar{D}\Phi_1 D\Phi_2\right). \quad (37)$$

5. Conclusions

In this work we have applied Galley's principle of stationary action to fermionic and supersymmetric systems. Except for the additional boundary condition, the generalization is straightforward. Examples discussed fit the prescription for a non conservative lagrangian. The generalized potential can be written in superfield form.

6. Acknowledgments

The authors thank VIEP-BUAP for financial support.

Appendix A. Superspace Supersymmetry

$N = 2$ superspace contains a commuting variable t, which plays the role of time, and two anticommuting variables θ and $\bar{\theta}$, wich are complex conjugate of each other [6]. Superspace translations are defined

$$\delta\theta = \bar{\epsilon}, \quad (A.1)$$
$$\delta\bar{\theta} = \epsilon, \quad (A.2)$$
$$\delta t = -i(\epsilon\bar{\theta} + \bar{\epsilon}\theta). \quad (A.3)$$

The variation of a supercoordinate $\Phi(t, \theta, \bar{\theta})$ is given by

$$\delta\Phi \equiv \epsilon Q\Phi + \bar{\epsilon}\bar{Q}\Phi, \quad (A.4)$$

where, the generators of super-translations are defined

$$Q = \partial_\theta - i\bar{\theta}\partial_t, \quad (A.5)$$
$$\bar{Q} = \partial_{\bar{\theta}} - i\theta\partial_t, \quad (A.6)$$

which satisfy the algebra of supersymmetry

$$\{Q, \bar{Q}\} = -2P, \quad (A.7)$$
$$QQ = \bar{Q}\bar{Q} = 0. \quad (A.8)$$

The covariant derivatives are

$$D = \partial_\theta + i\bar{\theta}\partial_t, \quad (A.9)$$
$$\bar{D} = \partial_{\bar{\theta}} + i\theta\partial_t, \quad (A.10)$$

which satisfy the algebra

$$\{D, D\} = 2P, \quad (A.11)$$
$$DD = \bar{D}\bar{D} = 0. \quad (A.12)$$

In order to get expressions in component form, we use the following expansion $\Phi(t, \theta, \bar{\theta}) = q(t) + i\theta\bar{\psi} + i\bar{\theta}\psi - \theta\bar{\theta}A$.

References

[1] Galley CR 2013. *Phys. Rev. Lett.* **110** 174301.
[2] Galley v, Tsang D, Stein LC 2104. arXiv:1412.3082 [math-ph].
[3] Schwinger J 1961, *J. Math. Phys.* **2** 407.
[4] Martínez-Pérez NE and Ramírez C 2015, *Supersymmetric non conservative systems*, arXiv:1501.05018.
[5] Teitelboim C, Galvao C.A.P. 1980 *J. Math. Phys.* **21** 1863-80.
[6] Bellucci S(Ed) 2006. *Supersymmetric Mechanics - Vol. 1. Supersymmetry, Non conmutativity and Matrix Models*. Lecture Notes in Physics. (Berlin: Springer) p 54.

GEOMETRIC DESCRIPTION OF BLACK HOLE THERMODYNAMICS WITH HOMOGENEOUS FUNDAMENTAL EQUATION

Alberto Sánchez

Departamento de Posgrado, CIIDET, API52, Quertaro, QRO 76000, MEXICO (May 26, 2015)

E-mail: `asanchez@ciidet.edu.mx`

Abstract. We investigate the Geometrothermodynamics of the 2-dimensional submanifold $\mathcal{E}$ corresponding the space of thermodynamic equilibrium states for a static and spherically symmetric hairy to black hole solution in Lorentz non-invariant massive gravity. We show that it is possible to consider a fundamental thermodynamic function as a homogeneous function of degree 1 in the extensive variables, as classical thermodynamics demand, in order to be consistent with the physical meaning of the intensive variables. The geometry of the space of equilibrium states is computed showing that it contains information about the thermodynamic interaction, critical points, and phase transitions structure.

1. Introduction

The research done by Hawking [1] and Bekenstein[2, 3, 4] in the early 1970s, showed that it is possible to find a relationship between the properties of the black holes (event horizon) and the laws of thermodynamics. For example, the surface gravity is related with the temperature, and the sum of the areas of the black hole horizons cannot decrease, which resembles the classical thermodynamic entropy. Therefore, an entropy which is proportional to the area was attributed to the horizon of the black hole. Thus, they also realized that one can formulate the four laws of black hole dynamics in a manner analogous to the laws of classical thermodynamics [5].

In spite of the fact that the main objection with respect to this analogy was resolved [6] (one that argued that if black holes possess entropy as well as energy, then they must have a non-zero temperature and must radiate which seemed to contradict the view that nothing can escape the black hole horizon) still some others remain, such as the non-homogeneity of the fundamental equation [7] which describe the thermodynamics of the black holes, or the meaning of a phase transition in these exotic thermodynamic systems.

The study of the thermodynamics of the black holes by means of geometry has been a subject of intensive research [8, 9, 10, 11, 12, 13]. This geometric study has been considered in several papers by means of different approaches like Weinhold's theory [14], Ruppeiner's theory [15] or the most recent theory called geometrothermodynamics [16]. Geometrothermodynamics (GTD) is a formalism that relates a contact structure of the phase space $\mathcal{T}$ with the metric structure on a special subspace of $\mathcal{T}$ called the space of equilibrium states $\mathcal{E}$.

On the other hand, the most successful cosmological model that is in agreement with the observational data implies the existence of a vacuum energy related with the cosmological constant Λ whose magnitude is unnatural from the effective field theory point of view [17]. Hence, a dark energy is needed to reconcile general relativity with the observations. The dark energy solves the problem of the accelerated expansion of the Universe. Nevertheless, the interest to explain the acceleration of the universe without resorting to the dark energy has motivated the search for large-distances modified theories of the gravity, the Lorentz- breaking massive gravity is one of these models which is free from pathologies such as ghosts, low strong coupling scales or instabilities at full non-perturbative level [18, 19, 20]. Although these models do not require the existence of Λ, the cosmological constant problem remains open [21]. A generalized Schwarzschild solution for this model has been obtained by D. Comelli et al. which is an exact black hole solution showing a nonanalytic hair [22].

In this work, we will use geometrotermodynamics to formulate an invariant geometric representation of the thermodynamics of a static and spherically symmetric hairy black hole solution in massive gravity, using a thermodynamic fundamental equation of degree 1.

This paper is organized as follows. In Section 2, we review the generalized Schwarzschild solution for massive gravity and its thermodynamics. In Sec. 3, we present a brief review of the GTD formalism. In Sec. 4, we apply the formalism of GTD to the generalized Schwarzschild solution in massive gravity and show that the geometric properties of the equilibrium space are in correspondence with thermodynamic properties of the black hole. Finally, in Sec. 5 we present the conclusions.

2. Thermodynamics of the black hole solution in Lorentz non-invariant massive gravity

The general action corresponding to massive gravity is given by the expression [20, 22, 23, 24, 25, 26]:

$$I = \int_{\mathcal{M}} d^4\sqrt{g}\Big[-\frac{1}{16\pi}R + \Lambda^4\mathcal{F}(X, Vi, W^{ij})\Big] - \int_{\partial\mathcal{M}} d^3\sqrt{\gamma}\frac{K}{8\pi}\,, \tag{1}$$

where $\mathcal{F}$ is a function of four scalar fields ϕ^μ that are minimally coupled to gravity by the covariant derivatives since:

$$X = \Lambda^{-4}g^{\mu\nu}\partial_\mu\phi^0\partial_\nu\phi^0\ ;\ V^i = \Lambda^{-4}\partial^\mu\phi^i\partial_\mu\phi^0\ ;\ W^{ij} = \Lambda^{-4}\partial^\mu\phi^i\partial_\mu\phi^j - \frac{V^iV^j}{X}\,, \tag{2}$$

with latin and greek indices running on the space and spacetime, respectively. The second integral is instead the Gibbons-Hawking-York boundary term [28, 29] where γ is the metric induced on the boundary $\partial\mathcal{M}$ and K is the trace of the extrinsic curvature $K_{ij} = \frac{1}{2}\gamma^k{}_i\nabla_k n_j$ of $\mathcal{M}$ with unit normal n^i. Such a boundary term is required to have a well-defined variational principle in the presence of the border $\partial\mathcal{M}$.

Spherically symmetric black holes in massive gravity have been investigated in [22, 28, 29] and it was found that a set of coordinates always can be found where the solution can be written in the form:

$$ds^2 = -f(r)dt^2 + f(r)^{-1}dr^2 + r^2(d\theta^2\sin^2\theta d\varphi^2)\,, \tag{3}$$

and the fields are given by the expressions [28]:

$$\phi^0 = \Lambda^2[t + h(r)]\ ;\ \phi^i = \phi(r)\frac{\Lambda^2 x^i}{r}\,. \tag{4}$$

The metric functions f, h and ϕ are given by the following equations,

$$f = 1 - \frac{2M}{r} - \frac{Q}{r^{\lambda}} \tag{5}$$

$$h = \pm \int \frac{dr}{f}\left[1 - f\Big(\frac{\lambda(\lambda-1)Q}{12m^2 r^{\lambda+2}} + 1\Big)^{-1}\right]^{\frac{1}{2}}, \tag{6}$$

$$\phi(r) = r\,, \tag{7}$$

where $\lambda \neq 1$ is a positive constant, M and Q are integration constants. In the case $\lambda > 1$ the gravitational potential is asymptotically Newtonian and the parameter M coincides with the ADM mass, while Q is a scalar charge whose presence reflects the modification of the gravitational interaction as compared to General Relativity.

The roots of the lapse function $f(g_{tt} = 0)$ define the horizons $r = r_{\pm}$ of the spacetime. In particular, the null hypersurface $r = r_+$ can be shown to correspond to an event horizon, which in this case is also a Killing horizon, whereas the inner horizon at r_- is a Cauchy horizon. Therefore from $f(r_+) = 0$ [26] we get,

$$1 - \frac{2M}{r_+} - \frac{Q}{r_+^{\lambda}} = 0\,, \tag{8}$$

From the expression (8) for the horizon radii r_+ we obtain the following relationship,

$$M(r_+, Q) = \frac{1}{2} r_+ \left[1 - \frac{Q}{r_+^{\lambda}}\right], \tag{9}$$

From the area-entropy relationship, $S = \pi r^2{}_+$, the equation (8) can be rewritten as

$$M(S,Q) = \frac{1}{2}\left(\frac{S}{\pi}\right)^{\frac{1}{2}}\left[1 - \frac{Q}{\left(\frac{S}{\pi}\right)^{\frac{\lambda}{2}}}\right]. \tag{10}$$

This equation relates all the thermodynamic variables entering the black hole metric in the form of a fundamental thermodynamic equation $M = M(S,Q)$. Also, this can be considered as the thermodynamic potential related to the canonical ensemble of the system so that the corresponding heat capacity contains information about the phase transitions of the system. Equation (10) is an inhomogeneous function in the extensive variables S and Q. Following Davies [30] we homogenisize the fundamental equation by redefining the parameters Q and S as,

$$Q = q^{\lambda}\,, \quad ; \quad S = s^2\,. \tag{11}$$

Then, equation (10) becomes a homogeneous function of degree 1 in the extensive variables. With the definitions (11) we have,

$$m(s,q) = \frac{s}{2\sqrt{\pi}}\left[1 - \frac{q^{\lambda}\pi^{\frac{\lambda}{2}}}{s^{\lambda}}\right]. \tag{12}$$

This procedure was performed explicitly in the context of GTD in [31]. Under this consideration the first law of the thermodynamics can be written as,

$$dm = Tds + \phi dq\,, \tag{13}$$

with $m = m(s, q)$. According with equation (13), the expression for the temperature T and the potential ϕ are given by the thermodynamic equilibrium conditions: $T = \left(\frac{\partial m}{\partial s}\right)_q$ and $\phi = \left(\frac{\partial m}{\partial q}\right)_s$, which lead to the following results,

$$T = \frac{1}{2\sqrt{\pi}}\left[1 - \frac{q^{\lambda}\pi^{\frac{\lambda}{2}}(1-\lambda)}{s^{\lambda}}\right], \tag{14}$$

$$\phi = -\frac{1}{2\pi^{\frac{1-\lambda}{2}}}\left[\frac{q}{s}\right]^{\lambda-1}. \tag{15}$$

It is easy to show that the temperature (14) coincides with the Hawking temperature. The heat capacity at constant values of q is given as

$$C_q = T\left(\frac{\partial s}{\partial T}\right)_q = \left(\frac{\frac{\partial M}{\partial s}}{\frac{\partial^2 M}{\partial s^2}}\right)_q, \tag{16}$$

where the subscript indicate that we should compute keeping the charge constant. Using the fundamental equation (12) we get,

$$C_q = \frac{s\left[s^{\lambda} - q^{\lambda}\pi^{\frac{\lambda}{2}}(1-\lambda)\right]}{q^{\lambda}\pi^{\frac{\lambda}{2}}(1-\lambda)}. \tag{17}$$

According to Davies [30], second order phase transitions take place at those points where the heat capacity diverges, i. e., for

$$q^{\lambda}\pi^{\frac{\lambda}{2}}(1-\lambda) = 0\,, \tag{18}$$

in this case, because $\lambda \neq 1$ there are no points where the heat capacity becomes singular; therefore, this thermodynamic system does not present second order phase transitions. We can also observe that the the thermodynamic variables T and ϕ are homogeneous functions of zero order, as in the standard thermodynamics. It is worth noticing in the case of more general black hole solutions of massive gravity, the investigation of the heat capacity leads to a non trivial phase transition structure [27]. In the present case, however, no phase transitions are found.

3. Brief review of geometrothermodynamics

Geometrothermodynamics (GTD) [16] is a formalism that has been applied to different thermodynamic systems(ordinary systems like the Van der Waals gas or exotic systems like black holes [32, 33, 34, 35, 36, 37, 38, 39] in order to get consistent results to describe geometrically the phases transitions and thermodynamic interaction using Legendre invariant metrics.

In GTD we work with a $(2n+1)$-dimensional manifold $\mathcal{T}$ with a set of coordinates Z^A which allow us to define a non-degenerate Legendre invariant metric G together with a linear differential 1-form Θ which fulfills the condition $\Theta \wedge (d\Theta)^n \neq 0$, where n is the number of thermodynamic degrees of freedom, $\wedge$ represents the exterior product and d the exterior derivative. In GTD we also have the space of thermodynamic equilibrium states which is a submanifold $\mathcal{E} \subset \mathcal{T}$ defined

Journal of Physics: Conference Series **654** (2015) 012006 doi:10.1088/1742-6596/654/1/012006

by means of a smooth embedding mapping $\varphi : \mathcal{E} \longrightarrow \mathcal{T}$ such that the pullback $\varphi^*(\Theta) = 0$. By means of $\varphi^*(G) = g$ a metric g is induced in $\mathcal{E}$, giving a Riemannian structure to this space. Therefore, in GTD the physical properties of a thermodynamic system in a state of equilibrium are described in terms of the geometric properties of the corresponding space $\mathcal{E}$.

If we consider the $(2n+1)$-dimensional space $\mathcal{T}$ coordinatized by the set $Z^A = \left\{\Phi\,, E^a\,, I^a\right\}$ where $A = 0, \ldots, 2n$ and $a = 1, \ldots n$, the 1-form Θ will be,

$$\Theta = d\Phi - I_a dE^a\,. \tag{19}$$

We choose now the subset E^a as coordinates of $\mathcal{E}$. Then the mapping φ is given by,

$$\varphi : (E^a) \longrightarrow (\Phi\,, E^a\,, I^a)\,, \tag{20}$$

and the condition,

$$\varphi^*(\Theta) = \varphi^*(d\Phi - \delta_{ab} I^a dE^b) = 0\,, \tag{21}$$

leads to the standard conditions of the thermodynamic equilibrium and the first law of thermodynamics,

$$\frac{\partial \Phi}{\partial E^a} = I_a\,, \qquad d\Phi = I_a dE^a\,. \tag{22}$$

The second law of the thermodynamics under this formalism is written as,

$$\frac{\partial^2 U}{\partial E^a \partial E^b} \geq 0\,, \quad ; \quad \frac{\partial^2 S}{\partial F^a \partial F^b} \leq 0\,, \tag{23}$$

where U and S represent the energy and entropy for each of the corresponding thermodynamic systems. Here E^a (F^a) represent all the extensive variables other than U (S).

In GTD the only requirement for defining a metric G of the space $\mathcal{T}$ is that it fulfills the condition of Legendre invariance. Therefore, we have many possibilities of constructing a metric with these features, one of them is the following,

$$G = \Theta^2 + (\delta_{ab} E^a I^b)(\eta_{cd} dE^c dI^d)\,, \tag{24}$$

where $\delta_{ab} = \text{diag}(1, 1, \ldots, 1)$ and $\eta_{ab} = \text{diag}(-1, 1, \ldots, 1)$. It can be shown that the metric (24) is invariant with respect to a total Legendre transformation which changes the variables without tilde $\{\Phi, E^a, I^a\}$ to the tilde variables $\{\tilde{\Phi}, \tilde{E^a}, \tilde{I^a}\}$ using the following algebraic rules,

$$\Phi = \tilde{\Phi} - \tilde{E}_a \tilde{I}^a\,, \qquad E^a = -\tilde{I}^a\,, \qquad I^a = \tilde{E}^a\,. \tag{25}$$

Applying the pullback φ^* to the metric (24), we obtain the corresponding thermodynamic metric g,

$$g^{GTD} = \varphi^*(G) = \left(E^c \frac{\partial \Phi}{\partial E^c}\right)\left(\eta_{ab}\delta^{bc} \frac{\partial^2 \Phi}{\partial E^c \partial E^d} dE^a dE^d\right)\,, \tag{26}$$

which depends only of the fundamental potential $\Phi = \Phi(E^a)$. If we know the fundamental potential of the thermodynamic system that we want to study, the corresponding metric g can be computed explicitly and the relations between thermodynamic and geometry can be also studied.

4. Geometrothermodynamics of the black hole solution in Lorentz non-invariant massive gravity

According with [40], the geometrothermodynamic metric that we should use to investigate the thermodynamic properties of black holes is given by the expression (26). If we define $\Phi = m$ and we assume that $E^a = \{s, q\}$, then we have a system with two thermodynamic degrees of freedom. Under these considerations we get,

$$g^{GTD} = \Big(s\frac{\partial m}{\partial s} + q\frac{\partial m}{\partial q}\Big)\Big(-\frac{\partial^2 m}{\partial s^2}ds^2 + \frac{\partial^2 m}{\partial q^2}dq^2\Big)\,. \tag{27}$$

Using the expressions for the m, as was given in Eq. (12), we obtain explicitly the GTD metric coefficients which can be written as,

$$g_{ss}^{GTD} = \frac{\lambda(\lambda-1)}{4\pi^{\frac{2-\lambda}{2}}}\frac{q^\lambda\big[q^\lambda\pi^{\frac{\lambda}{2}} - s^\lambda\big]}{s^{2\lambda}}\,, \tag{28}$$

$$g_{qq}^{GTD} = -\frac{\lambda(\lambda-1)}{4\pi^{\frac{2-\lambda}{2}}}\frac{q^{\lambda-2}\big[q^\lambda\pi^{\frac{\lambda}{2}} - s^\lambda\big]}{s^{2\lambda-2}}\,, \tag{29}$$

and the equation (27) takes the form,

$$g^{GTD} = f(s,q)\Big[\frac{ds^2}{s^2} - \frac{dq^2}{q^2}\Big]\,, \tag{30}$$

where,

$$f(s,q) = \frac{\lambda(\lambda-1)}{4\pi^{\frac{2-\lambda}{2}}}\frac{q^\lambda\big[q^\lambda\pi^{\frac{\lambda}{2}} - s^\lambda\big]}{s^{2\lambda-2}}\,. \tag{31}$$

The curvature scalar corresponding to the metric (30) takes the form,

$$R^{GTD} = 0\,. \tag{32}$$

This result tell us that the curvature scalar of the space of thermodynamic equilibrium states does not have points where becomes singular. Also, in accordance with the geometrothermodynamics, the system does not have interaction thermodynamics [40].

5. Conclusions

In this work, we studied the geometrothermodynamics of a black hole solution in massive gravity. Using this formalism we found the geometric properties of the corresponding manifold of equilibrium states. We found that the corresponding thermodynamic curvature turned out to be zero, indicating that there is not thermodynamics interaction. We also show that the GTD correctly describes the fact that this thermodynamic system has not phase transitions, since they are characterized by divergencies of the heat capacity which are described in GTD by curvature singularities.

It is interesting that, according to GTD, the black hole considered in this work does not posses thermodynamic interaction. In fact, this intriguing behavior has been also found in the case of topological black holes in Horava-Lifshitz gravity [41] and in cosmological models [42]. This means that in GTD the presence of gravitational interaction does not necessarily imply the presence of thermodynamic interaction. We expect to study this intriguing relationship in future works.

We assumed in this work Davies's proposal to solve the problem of the lack of homogeneity of the fundamental equation. We have considered the definition (11), with which the fundamental equation (9) becomes a homogeneous function of degree 1 in the extensive variables in order to have consistency with the physical meaning of the intensive variables which must be homogenous of degree zero.

Acknowledgements
This work was supported by Conacyt-Mexico, Grant No. 166391 and DGAPA-UNAM, Grant No. 106110.

References

[1] Hawking S W 1971 *Phy.Rev.Lett* **26** 1344
[2] Bekenstein J D 1972 *Nuovo Cim. Lett.* **4** 737740
[3] Bekenstein J D 1973 *Phys. Rev.* D **7** 23332346
[4] Bekenstein J D 1974 *Phys. Rev.* D **9** 32923300.
[5] Bardeen J M, Carter B and Hawking S W 1973 *Commun. Math. Phys.* **31** 161170
[6] Hawking S W 1975 *Commun. Math. Phys.* **43** 199220
[7] Callen H B 1981 *Thermodinamics* (John Wiley and Sons, Inc., New York), pag. 25
[8] Bravetti A, Momeni D, Myrzakulov R and Quevedo H 2013 Geometrothermodynamics of higher dimensional black holes *General Relativity and Gravitation* **45** N8 pp 1603-1617
[9] Aman J E, Bengtsson I and Pidokrajt N 2003 Geometry of black hole thermodynamics *Gen. Rel. Grav.* **35** 1733
[10] Aman J E and Pidokrajt N 2006 Geometry of higher-dimensional black hole thermodynamics *Phys. Rev.* D **73** 024017
[11] Aman J E and Pidokrajt N 2006 Flat Information Geometries in Black Hole Thermodynamics *Gen. Rel. Grav.* **38** 1305
[12] Shen J, Cai R G, Wang B and Su R K 2007 Thermodynamic Geometry and Critical Behavior of Black Holes *Int.J.Mod.Phys.* **A22:** 11-27
[13] Cai R G and Cho J H 1999 Thermodynamic curvature of the BTZ black hole *Phys. Rev.* D **60** 067502
[14] Weinhold F 1975 Metric geometry of equilibrium thermodynamics. III. Elementary formal structure of a vector-algebraic representation of equilibrium thermodynamics., J. Chem. Phys. **63** 2488
[15] Ruppeiner G 1979 Thermodynamics: A Riemannian geometric model *Phys. Rev.* A **20** 1608
[16] Quevedo H 2007 Geometrothermodynamics *J. Math. Phys.* **48** 013506
[17] Copeland E J, Sami M and Tsujikawa S 2006 Dynamics of dark energy *Int. J. Mod. Phys.* **D15** 17531936
[18] Arkani-Hamed N, Georgi H and Schwartz MD 2003 Effective Field Theory for Massive Gravitons and Gravity in Theory Space *Annals Phys.* **96** 305
[19] Arkani-Hamed N, Cheng H C, Luty M A and Mukohyama S 2004 Ghost Condensation and a Consistent Infrared Modification of Gravity *JHEP* **074** 0405
[20] Dubovsky S L 2004 Phases of massive gravity *JHEP* **076** 0410
[21] Weinberg S 1989 the cosmological constant problem *Rev. Mod. Phys.* **61** 123
[22] Comelli D, Nesti F and Pilo L 2011 Stars and (Furry) Black Holes in Lorentz Breaking Massive Gravity *Phys. Rev.* D **83** 084042
[23] Blas D 2008 Aspects of Infrared Modifications of Gravity *Preprint* arXiv:0809.3744vl [hep-th]
[24] Rubakov V A 2004 Lorentz-violating graviton masses: getting around ghosts, low strong coupling scale and VDVZ discontinuity *Preprint* hep-th/0407104
[25] Blas D, Comelli D, Nesti F and Pilo L 2009 Lorentz-breaking massive gravity in curved space *Phys. Rev.* D **80** 044025
[26] Capela F and Nardini G 2012 Hairy Black Holes in Massive Gravity; Thermodynamics and Phase Structure *Phys. Rev.* D **86** 024030

[27] S. Hendi H, Panahiyan S, Panah B E and Momennia M 2015 Geometrical thermodynamics of phase transition: charged black holes in massive gravity *Preprint* 1506.07262v2 [hep-th]
[28] Bebronne M V and Tinyakov P G 2011 Black hole solutions in massive gravity *JHEP* **100** 0904 [Erratum-ibid. 1106 018] *Preprint* arXiv:0902.3899 [gr- qc]
[29] Berezhiani Z, Comelli D, Nesti F and Pilo L 2008 Exact Spherically Symmetric Solutions in Massive Gravity *JHEP* **130** 0807
[30] Davies P C W 1977 The Thermodynamic Theory of Black Holes *Proc. Roy. Soc. Lond.* **A** 353 499
[31] Azieg-Ainou M 2014 Geometrothermodynamics: Comments, critics, and supports *Preprint* arXiv:1311.6963 [gr-qc]
[32] Quevedo H 2008 Geometrothermodynamics of black holes *Gen. Rel. Grav.* **40** 971
[33] Quevedo H, Sánchez A, Taj S and Vázquez A 2011 Phase transitions in geometrothermodynamics *Gen.Rel.Grav.* **43** 1153-1165
[34] Alvarez J L, Quevedo H and Sánchez, A 2008 Unified geometric description of black hole thermodynamics.., *Phys. Rev.* D **77** 084004
[35] Quevedo H and Ramiírez A 2012 A geometric approach to the thermodynamics of the van der Waals system *Preprint* arXiv:1205.3544vl [math-ph]
[36] Quevedo H, Sánchez A and Vázquez A 2010 Curvature as a Measure of the Thermodynamic Interaction *Journal of the Korean Physical Society* **57** 646-650
[37] Quevedo H and Sánchez A 2008 Geometrothermodynamics of asymptotically Anti-de Sitter black holes *JHEP* **09** 034
[38] Quevedo H and Sánchez A 2009 Geometric description of BTZ black hole thermodynamics *Phys. Rev.* D **79** 024012
[39] Quevedo H and Sánchez A 2009 Geometrothermodynamics of black holes in two dimensions *Phys. Rev.* D. **79** 087504
[40] Quevedo H, Sánchez A, Taj S and Vázquez A 2011 Phase Transitions in Geometrothermodynamics *Gen. Relativ. Gravit.* **43** 1153
[41] Quevedo H, Sánchez A, Taj S and Vzquez A, Geometrothermodynamics in Horava–Lifshitz gravity 2011 *Gen. Relativ. Gravit.* **43**, 1153
[42] Aviles A, Bastarrachea–Almodovar A, Campuzano L and Quevedo H 2012 Extending the generalized Chaplygin gas model by using geometrothermodynamics *Phys. Rev.* D **86** 063508

Journal of Physics: Conference Series **654** (2015) 012007 doi:10.1088/1742-6596/654/1/012007

Time-varying cosmological term

J. Socorro[1], M. D'oleire[1] and Luis O. Pimentel[2]

[1]Departamento de Física, DCeI, Universidad de Guanajuato-Campus León, C.P. 37150, León, Guanajuato, México

[2] Departamento de Física, Universidad Autónoma Metropolitana, Apartado Postal 55-534, C.P. 09340 México, DF, México

E-mail: socorro@fisica.ugto.mx

E-mail: marcodoleire@gmail.com

E-mail: lopr@xanum.uam.mx

Abstract. We present the case of time-varying cosmological term using the Lagrangian formalism characterized by a scalar field ϕ with standard kinetic energy and arbitrary potential $V(\phi)$. This model is applied to Friedmann-Robertson-Walker (FRW)cosmology. Exact solutions of the field equations are obtained by a special ansats to solve the Einstein-Klein-Gordon equation and a particular potential for the scalar field and barotropic perfect fluid. We present the evolution on this cosmological term with different scenarios.

1. introduction

The present phase of an accelerated expansion of the universe stands as one of the most challenging open problems in modern cosmology and astrophysics. This acceleration is characterized by which is popularly known as dark energy. Among many possible alternatives, the simplest candidate for dark energy is the vacuum energy which is mathematically equivalent to the cosmological constant. Models with different decay laws for the variation of the cosmological term were investigated during last two decades in non covariant way, [1, 2, 3, 4, 5, 6, 7, 8, 9], in particular, in [10] appear several evolution relations for Λ which many author have used. Anisotropic cosmological models, also has been treated in this formalism from different points of view [11, 12, 13, 14, 15, 16, 17, 18, 19, 20, 21, 22].

In this work we present an analysis in covariant way, using the Lagrangian density of the standard scalar field. The main idea arises by proposing that in the cosmological constant case, the scalar potential is identified as $\mathrm{V}(\phi) = 2\Lambda$, when Λ is a constant. So, we generalized this idea and suggest that this correspondence is valid yet when this cosmological term has a temporal dependence, i.e., $\mathrm{V}(\phi(\mathrm{t})) = 2\Lambda(\mathrm{t})$. We include a barotropic equation state between the pressure and energy density of the scalar field, $\mathrm{p}_\phi = \omega_\phi \rho_\phi$, quantities that we shall define in the following lines. In order to develop the study presented here, we consider a proportionality between the energy density of the scalar field and the energy density of the barotropic perfect fluid, with the proportionality constant m_ϕ, that is, $\rho_\phi = \mathrm{m}_\phi \rho$.

The corresponding Lagrangian density including a scalar field is

$$\mathcal{L}[\mathrm{g}, \phi] = \sqrt{-\mathrm{g}} \left(\mathrm{R} - \frac{1}{2} \mathrm{g}^{\mu\nu} \nabla_\mu \phi \nabla_\nu \phi + \mathrm{V}(\phi) \right) + \sqrt{-\mathrm{g}} \mathcal{L}_{\mathrm{matter}} \tag{1}$$

where R is the Ricci scalar, $\mathcal{L}_{matter}$ correspond to barotropic perfect fluid, $p = \omega\rho$, ρ is the energy density and p is the pressure of the fluid in co-moving frame and ω is constant.

The corresponding variation of (1), with respect to the metric and the scalar field yields to the Einstein and Klein-Gordon equations, respectively

$$R_{\alpha\beta} - \frac{1}{2}g_{\alpha\beta}R = -\frac{1}{2}\left(\nabla_\alpha\phi\nabla_\beta\phi - \frac{1}{2}g_{\alpha\beta}g_{\mu\nu}\nabla_\mu\phi\nabla_\nu\phi\right) + \frac{1}{2}g_{\alpha\beta}V(\phi) - 8\pi G T_{\alpha\beta}, \tag{2}$$

$$\Box\phi - \frac{\partial V}{\partial\phi} = 0. \tag{3}$$

From (2) it could be deduced that the energy-momentum tensor associated with the scalar field is

$$T^{(\phi)}_{\alpha\beta} = \frac{1}{2}\left(\nabla_\alpha\phi\nabla_\beta\phi - \frac{1}{2}g_{\alpha\beta}g_{\mu\nu}\nabla_\mu\phi\nabla_\nu\phi\right) - \frac{1}{2}g_{\alpha\beta}V(\phi) \tag{4}$$

and the corresponding tensor for a barotropic perfect fluid is

$$T_{\alpha\beta} = (p+\rho)\,u_\alpha u_\beta + g_{\alpha\beta}p$$

here u_α is the four-velocity in comoving frame. The line element to be considered in this work is the FRW one,

$$\begin{aligned} ds^2 &= -N^2(t)dt^2 + A^2(t)\left[\frac{dr^2}{1-\kappa r^2} + r^2(d\theta^2 + \sin^2\theta d\phi^2)\right] \\ &= -d\tau^2 + A^2(\tau)\left[\frac{dr^2}{1-\kappa r^2} + r^2(d\theta^2 + \sin^2\theta d\phi^2)\right], \end{aligned} \tag{5}$$

where we identify the time transformation $N(t)dt = d\tau$, this transformation will be used in the whole work, and in special gauge we recover directly the cosmic time t.

2. field equations

Making use of metric (5) and a comoving fluid, equations (2) y (3) are now

$$\frac{3A'^2}{A^2} + \frac{3\kappa}{A^2} - 8\pi G\rho - \frac{1}{4}\phi'^2 - \frac{V(\phi)}{2} = 0, \tag{6}$$

$$\frac{2A''}{A} + \frac{A'^2}{A^2} + \frac{\kappa}{A^2} + 8\pi Gp + \frac{1}{4}\phi'^2 - \frac{1}{2}V(\phi) = 0, \tag{7}$$

$$\left[3\frac{A'}{A}\phi'^2 + \phi'\phi''\right] = -V', \tag{8}$$

the Klein-Gordon (8) can be rewritten as

$$\frac{d}{d\tau}\left[\mathrm{Ln}\left(A^6\frac{\phi'^2}{2}\right)\right] = -\frac{V'}{\frac{\phi'^2}{2}}. \tag{9}$$

In the literature there are some articles where the authors try to solve these field equations in general way, for instance, in reference [23], the authors present an elaborated technique to solve the Klein-Gordon equation (8), and in [24] they use an algebraic method to obtain exact solutions taking as the basic variable the energy density of the scalar field.

In order to solve this set of equations, we introduce the ansatz, of considering that the energy density of the field ϕ is proportional to the energy density of the barotropic perfect fluid,

$\rho_\phi = m_\phi \rho$, where m_ϕ is a *positive constant.* The scaling behavior occurs when $\mathrm{m}_\phi < 1$, otherwise, the quintessence field is dominant.

The energy density and pressure of the field ϕ are given as

$$16\pi \mathrm{G}\rho_\phi = \frac{1}{2}\phi'^2 + \mathrm{V}(\phi), \qquad 16\pi \mathrm{G}\mathrm{p}_\phi = \frac{1}{2}\phi'^2 - \mathrm{V}(\phi)$$

now, equations (6,7) are rewritten as

$$\frac{3\mathrm{A}'^2}{\mathrm{A}^2} + \frac{3\kappa}{\mathrm{A}^2} - 8\pi\mathrm{G}\left(\rho + \rho_\phi\right) = 0, \tag{10}$$

$$\frac{2\mathrm{A}''}{\mathrm{A}} + \frac{\mathrm{A}'^2}{\mathrm{A}^2} + \frac{\kappa}{\mathrm{A}^2} + 8\pi\mathrm{G}\left(\mathrm{p} + \mathrm{p}_\phi\right) = 0. \tag{11}$$

We will make now the assumption that the scalar field is a barotropic fluid: $\mathrm{p}_\phi = \omega_\phi \rho_\phi$, where ω_ϕ is a constant that play the same role of the ω parameter in the barotropic perfect fluid. Under this proposal, the field equations are

$$\frac{3\mathrm{A}'^2}{\mathrm{A}^2} + \frac{3\kappa}{\mathrm{A}^2} - 8\pi\mathrm{G}\rho_\mathrm{T} = 0, \tag{12}$$

$$\frac{2\mathrm{A}''}{\mathrm{A}} + \frac{\mathrm{A}'^2}{\mathrm{A}^2} + \frac{\kappa}{\mathrm{A}^2} + 8\pi\mathrm{G}\mathrm{p}_\mathrm{T} = 0, \tag{13}$$

$$\frac{\mathrm{d}}{\mathrm{d}\tau}\left[\mathrm{Ln}\left(\mathrm{A}^6 \frac{\phi'^2}{2}\right)\right] = -\frac{\mathrm{V}'}{\frac{\phi'^2}{2}}, \tag{14}$$

where the energy density $\rho_T = \rho + \rho_\phi = \alpha_\phi \rho$ with $\alpha_\phi = 1 + m_\phi > 1$, and the pressure $p_T = p + p_\phi = \omega_T \rho_T$ and the last barotropic index is given by

$$\omega_\mathrm{T} = \left(\frac{\omega + \mathrm{m}_\phi \omega_\phi}{1 + \mathrm{m}_\phi}\right). \tag{15}$$

So far we have three barotropic indices, two assumed, ω and ω_ϕ, and one implied, ω_T. We show now that all three must be the same. Notice that he first two equations are the standard FRW equation for the total fluid with barotropic index ω_T. Also notice that under the assumption of the relation between the density of the scalar field and the density of the perfect fluid, and the proportionality of the p_ϕ with ρ_ϕ then the potential V, $\phi'^2/2$ and all the densities $(\rho, \rho_\phi, \rho_T)$are proportional to each other. From Eqs. (12,13) we now that

$$\rho_\mathrm{T} = \mathrm{c}_\mathrm{T} \mathrm{A}^{-3(\omega_\mathrm{T}+1)}. \tag{16}$$

Equation (14) can be written as

$$\frac{\mathrm{d}}{\mathrm{d}\tau}\left[\mathrm{Ln}\left(\mathrm{A}^6 \mathrm{V}^{\frac{2}{1+\omega_\phi}}\right)\right] = 0, \tag{17}$$

and the solution is

$$\mathrm{V}(\tau) = \mathrm{c}_\omega \, \mathrm{A}^{-3(1+\omega_\phi)}. \tag{18}$$

Then as a consequence of the proportionality between V and ρ_T the exponents in Eqs. (16, 18) should be equal,

$$\omega_\mathrm{T} = \omega_\phi, \qquad \Rightarrow \qquad \omega = \omega_\phi = \omega_\mathrm{T}. \tag{19}$$

3. General solution for flat space

In this section we present solutions to the field equations for the flat case. Equation (14) is written as

$$\frac{\mathrm{d}}{\mathrm{d}\tau}\left[\mathrm{Ln}\left(\mathrm{A}^6\mathrm{V}^{\frac{2}{1+\omega}}\right)\right]=0, \qquad \Rightarrow \qquad \mathrm{V}(\tau)=\mathrm{c}_\omega\,\mathrm{A}^{-3(1+\omega)}, \tag{20}$$

using the well known time evolution of the scalar factor for the barotropic fluid, reported in different places, in particular in reference [25], that for future convenience we write as

$$\mathrm{A}_\omega(\tau)=\begin{cases}[\mathrm{a}_\omega\,\tau]^{\frac{2}{3(\omega+1)}}, & \mathrm{a}_\omega=(\omega+1)\sqrt{6\pi G\alpha_\phi \mathrm{M}_\omega}, \qquad \omega\neq-1\\ \mathrm{e}^{2\sqrt{\frac{2}{3}\pi G\alpha_\phi \mathrm{M}_{-1}}\tau} & \qquad \omega=-1\end{cases}. \tag{21}$$

Therefore, the temporal dependence of the potential and the scalar field are

$$\mathrm{V}(\tau)=\mathrm{c}_\omega\frac{1}{(\mathrm{a}_\omega\tau)^2}, \qquad \Rightarrow \qquad \Delta\phi=\ell_\omega\,\mathrm{Ln}(\tau), \tag{22}$$

where the constants value c_ω and ℓ_ω are determined after substitution into the complete set of Einstein equation, with the scale factor solution (21), being the general solutions for any $\omega\neq\pm1$ the following relations are obtained

$$\mathrm{V}_\omega(\tau) = \frac{2\mathrm{m}_\phi(1-\omega)}{3(1+\omega)^2(1+\mathrm{m}_\phi)}\frac{1}{\tau^2}, \qquad \Leftrightarrow \qquad \Lambda(\tau)=\frac{\mathrm{m}_\phi(1-\omega)}{3(1+\omega)^2(1+\mathrm{m}_\phi)}\frac{1}{\tau^2}, \tag{23}$$

$$\Delta\phi(\tau) = \sqrt{\frac{4\mathrm{m}_\phi}{3(1+\omega)(1+\mathrm{m}_\phi)}}\,\mathrm{Ln}(\tau), \tag{24}$$

$$\mathrm{V}(\phi) = \frac{2\mathrm{m}_\phi(1-\omega)}{3(1+\omega)^2(1+\mathrm{m}_\phi)}\,\mathrm{e}^{-\sqrt{3(1+\omega)(1+\frac{1}{\mathrm{m}_\phi})}\Delta\phi}, \qquad \Leftrightarrow \qquad \Lambda(\phi)=\frac{\mathrm{m}_\phi(1-\omega)}{3(1+\omega)^2(1+\mathrm{m}_\phi)}\,\mathrm{e}^{-\sqrt{3(1+\omega)(1+\frac{1}{\mathrm{m}_\phi})}\Delta\phi} \tag{25}$$

We observe here that for all values of the barotropic parameter we have a decreasing cosmological function in time (remember the relation between the potential energy and the cosmological term, $\mathrm{V}(\tau)=2\Lambda(\tau)$), and that as function of the scalar field we have an exponential. In what follows we consider particular cases, divided into two branches of the barotropic parameter $0\leq\omega<1$, and $-1<\omega<0$.

3.1. Positive branch: $0\leq\omega<1$.

The relation between the kinetic term and the potential energy of the scalar field is

$$\frac{1}{2}\phi'^2=\frac{1+\omega}{1-\omega}V(\phi) \tag{26}$$

and we have the following particular case for the barotropic parameter that are of interest in cosmology and astrophysics.

(i) Dust scenario, $\omega=0$.
The set of equations (23,24,25) gives

$$\mathrm{V}_\omega(\tau) = \frac{2\mathrm{m}_\phi}{3(1+\mathrm{m}_\phi)}\frac{1}{\tau^2}, \tag{27}$$

$$\Delta\phi(\tau) = \sqrt{\frac{4\mathrm{m}_\phi}{3(1+\mathrm{m}_\phi)}}\,\mathrm{Ln}(\tau). \tag{28}$$

$$\mathrm{V}(\phi) = \frac{2\mathrm{m}_\phi}{3(1+\mathrm{m}_\phi)}\,\mathrm{e}^{-\sqrt{3\left(1+\frac{1}{\mathrm{m}_\phi}\right)}\Delta\phi}. \tag{29}$$

this behavior emerges when the universe have the following scale factor

$$\mathrm{A}_0(\tau) = [\mathrm{a}_0\,\tau]^{\frac{2}{3}}, \quad \mathrm{a}_0 = \sqrt{6\pi \mathrm{G}\alpha_\phi \mathrm{M}_0}. \tag{30}$$

Also this behavior is found using dynamical system and fitting that one critical point will be an attractor, obtaining that the corresponding factor λ in the exponential function $\mathrm{V}(\phi) \approx \mathrm{e}^{\lambda\phi}$, will satisfy $\lambda < -\sqrt{3}$ [26]. This value for the parameter λ was found using others techniques, in quantum solution and in supersymmetric quantum solutions in quantum cosmology, for the same flat FRW cosmological model [27, 28]. Chimento and Jakubi [23] found for inflationary era, solving the Einstein field equations with a power law expansion, that $\lambda = -\sqrt{2}$.

Is common say that when the scalar potential have a exponential behavior as in this case, the scale factor of the universe must have a fast grow in time. Remembering that $\alpha_\phi = 1 + m_\phi$, we take as base that the percent to usual matter becomes as 4%, considering the dark matter scenario, we need to order of 23%, so the approximate value to α_ϕ is 6. In the dark energy scenario, we need that α_ϕ near to 18. In these two last cases, the scale factor have a fast growing.

(ii) Radiation, $\omega = \frac{1}{3}$
The set of equations (23,24,25), for this value of the barotropic parameter, have the following form

$$\begin{aligned} \mathrm{V}_\omega(\tau) &= \frac{\mathrm{m}_\phi}{4(1+\mathrm{m}_\phi)}\frac{1}{\tau^2}, & (31)\\ \Delta\phi(\tau) &= \sqrt{\frac{\mathrm{m}_\phi}{1+\mathrm{m}_\phi}}\,\mathrm{Ln}(\tau), & (32)\\ \mathrm{V}(\phi) &= \frac{\mathrm{m}_\phi}{4(1+\mathrm{m}_\phi)}\,\mathrm{e}^{-2\sqrt{1+\frac{1}{\mathrm{m}_\phi}}\Delta\phi}. & (33) \end{aligned}$$

and the corresponding scale factor

$$\mathrm{A}_{\frac{1}{3}}(\tau) = \left[\mathrm{a}_{\frac{1}{3}}\,\tau\right]^{\frac{1}{2}}, \quad \mathrm{a}_{\frac{1}{3}} = \frac{4}{3}\sqrt{6\pi \mathrm{G}\alpha_\phi \mathrm{M}_{\frac{1}{3}}}.$$

The temporal dependence of the cosmological term goes to $\frac{1}{\tau^2}$, result reported in all references that used an analogue relation between the energy density in the scalar field and the energy density to the barotropic perfect fluid, in non covariant theory.

3.2. Negative branch: $-1 < \omega < 0$.
In this case we write the relation between the field pressure and density as

$$p_\phi = -|\omega|\rho_\phi; \rightarrow \frac{1}{2}\phi'^2 = \beta_\omega V(\phi), \qquad \beta_\omega = \frac{1-|\omega_\phi|}{1+|\omega_\phi|} \tag{34}$$

and we consider two particular values of ω

(i) For instance, when we choose the case $\omega_\phi = -\frac{2}{3}$, i.e., $|\omega_\phi| = \frac{2}{3}$,
The set of equations (23,24,25) have the following form

$$\begin{aligned} \mathrm{V}_\omega(\tau) &= \frac{10\mathrm{m}_\phi}{1+\mathrm{m}_\phi}\frac{1}{\tau^2}, & (35)\\ \Delta\phi(\tau) &= 2\sqrt{\frac{\mathrm{m}_\phi}{1+\mathrm{m}_\phi}}\,\mathrm{Ln}(\tau), & (36)\\ \mathrm{V}(\phi) &= \frac{10\mathrm{m}_\phi}{1+\mathrm{m}_\phi}\mathrm{e}^{-\sqrt{1+\frac{1}{\mathrm{m}_\phi}}\Delta\phi}, & (37) \end{aligned}$$

with a quadratic time dependence for the scale factor

$$A_{-\frac{2}{3}}(\tau) = \left[a_{-\frac{2}{3}}\right]^2 \tau^2, \qquad a_{-\frac{2}{3}} = \frac{1}{3}\sqrt{6\pi G\alpha_\phi M_{-\frac{2}{3}}}. \tag{38}$$

We consider that in this phenomenon scenario, the values of the m_ϕ yield between $(0, 1]$, for instance when $m_\phi = 1$, we recover the law to the potential field given by Chimento and Jakubi [23] with the corresponding scale factor.

(ii) When we choose $\omega = -\frac{1}{3}$, i.e., $|\omega_\phi| = \frac{1}{3}$, we have

$$V(\tau) = \frac{2m_\phi}{1 + m_\phi}\frac{1}{\tau^2}, \tag{39}$$

so, the scalar field is

$$\Delta\phi = \sqrt{\frac{2m_\phi}{1 + m_\phi}}\,\mathrm{Ln}(\tau), \tag{40}$$

thus, we can write $V(\phi)$

$$V(\phi) = \frac{2m_\phi}{1 + m_\phi} e^{-\sqrt{2\left(1+\frac{1}{m_\phi}\right)}\Delta\phi}, \tag{41}$$

with a linear evolution for the scale factor

$$A_{-\frac{1}{3}}(\tau) = a_{-\frac{1}{3}}\,\tau, \qquad a_{-\frac{1}{3}} = \frac{2}{3}\sqrt{6\pi G\alpha_\phi M_{-\frac{1}{3}}}. \tag{42}$$

4. Conclusions

In this work we have characterized the cosmological term $\Lambda(\tau)$ as proportional the potential for the scalar field. Assuming a proportionality between the energy density of the scalar field and the the density of a barotropic fluid of the matter content, an also assuming that the pressure and density of the scalar field satisfy a barotropic law, so that the field equation in the case of the FRW metric reduces to the standard cosmology in term of a total energy density and pressure that also satisfy a barotropic law. We found that for consistency all the different barotropic parameters should be the same. In the case of flat space we were able to find general exact solutions. A common characteristic of all the solutions is that the dynamic cosmological "constant" is decreasing in time as $\frac{1}{\tau^2}$ and that it is an exponential function of the scalar field. We also found the exponential behavior in the scalar field in the evolution of the universe, signal that the universe must have a fast growing scale factor. This fast growing correspond when we include the $m_\phi > 1$ parameter in the solution, that corresponds when the quintessence field is dominate in the universe, by instance we claim that if the percent to usual matter becomes as 4%, considering the dark matter scenario, we need to order of 23%, so the approximate value to α_ϕ is 6. In the dark energy scenario, we need that α_ϕ near to 18. In these two last cases, the scale factor have a fast growing. However the case $m_\phi < 1$ corresponds to scaling behavior with the usual matter. In all epochs analyzed in this work using the relation between the energy density to scalar field and the energy density of the ordinary matter, the behavior to the cosmological term goes as $\frac{1}{\tau^2}$, these results were found by other authors in non covariant theory [1, 2, 3, 4, 5, 6, 7, 8, 9, 10]. We consider that this behavior is dependent on the relation between the energy densities assumed in this work and others.

Acknowledgments

This work was partially supported by CONACYT 167335, 179881 grants. PROMEP grants UGTO-CA-3 and UAM-I-43. This work is part of the collaboration within the Instituto Avanzado de Cosmología and Red PROMEP: Gravitation and Mathematical Physics under project *Quantum aspects of gravity in cosmological models, phenomenology and geometry of space-time.* Many calculations where done by Symbolic Program REDUCE 3.8.

[1] W. Chen and Y.S. Wu, Phys. Rev. D **41**, 695 (1990).
[2] A-M. M- Abdel-Rahman, Gen. Rel. Grav. **22**, 655 (1990).
[3] D. Pavon, Phys. Rev. D bf 43, 375 (1991).
[4] J.C. Carvalho, J.A.C. Lima and I. Waga, Phys. Rev. D **46**, 2404 (1992).
[5] D. Kalligas, P. Wesson and C.W.F. Everitt, Gen. Rel. Grav.**24**, 351 (1992).
[6] J.A.S. Lima and J.M.F. Maia, Phys. Rev. D **49**, 5597 (1994).
[7] A.I. Arbab and A.M.M. Abdel-Rahaman, Phys. Rev. D **50**, 7725 (1994).
[8] J.A.S. Lima and J.C. Carvalho, Gen. Rel. Grav. **26**, 909 (1994).
[9] J.A.S. Lima and M. Trodden, Phys. Rev. D **53**, 4280 (1996).
[10] P.I. Fomin, P.A. Nakaznoy and S.I. Vilchinski, [gr-qc/0509042], *The closed Universe model with variable* Λ *term.*
[11] Aroonkumar Beesham, Phys. Rev. D **48**, 3539 (1993).
[12] Aroonkumar Beesham, Gen. Rel. Grav. **26**, 159 (1994).
[13] A.I. Arbab, Gen. Rel. Grav. **29**, 61 (1997).
[14] T. Singh, A. Beesham and W.S. Mbokazi, Gen. Rel. Grav. **30**, 573 (1998).
[15] A. Pradhan and A. Kumar, Int. J. of Mod. Phys. D **10**, 291 (2001).
[16] A. Pradhan, Int. J. of Mod. Phys. D **12**, 941 (2003).
[17] A. Pradhan and Om Prakash Pandey, Int. J. of Mod. Phys. D **12**, 1299 (2003).
[18] S. Carneiro, Int. J. of Mod. Phys. A **20**, 2465 (2005).
[19] A. Pradhan and P. Pandey, Astrophys. and Spa. Sci. **301**, 127 (2006).
[20] A. Pradhan, P. Pandey, K. Jotania and M. Kumar Yadav, Int. J. of Theor. Phys. **46**, 2774 (2007).
[21] A. Pradhan, Fizika B **16**, 205 (2007).
[22] G. Esposito, C. Rubano and P. Scudello, Class. Quantum Grav. **24**, 6255 (2007).
[23] L.P. Chimento and A.S. Jakubi, Int. J. of Mod. Phys D **5**, 71 (1996).
[24] M.A. Reyes, [arXiv:0806.2292], *On exact solutions to the scalar field equations in standard cosmology.*
[25] S.R. Berbena, Aarón V.B. Arellano, J. Socorro and Luis O. Pimentel. Rev. Mex. Fís. S **53** (2) 115 (2007), *The Einstein-Hamilton-Jacobi equation: searching the classical solution for barotropic FRW.*
[26] C. Hernández-Aguayo and L. A. Ureña-López, AIP Conf. Proc. **1473**, 68 (2011); *General dynamics of quintessence fields: Comparison with type Ia SNe.*
[27] J. Socorro and Marco D'oleire, Phys. Rev. D. **82**(4) 044008 (2010), *Inflation from supersymmetric quantum cosmology.*
[28] J. Socorro, M. Sabido, W. Guzmán, M. A. Granados, Capítulo cuatro, book: Procesos no lineales en la ciencia y la sociedad, (2013), Edit. Notabilis Scientia, Compilador M. A. Granados; ISBN: 978-607-8289-41-7, *Inflación Cosmológica Vista desde la Mecánica Cuántica Supersimétrica*

Hydrodynamic representation of the Klein-Gordon-Einstein equations in the weak field limit

Abril Suárez and Pierre-Henri Chavanis

Laboratoire de Physique Théorique, Université Paul Sabatier, 118 route de Narbonne 31062 Toulouse, France

E-mail: suarez@irsamc.ups-tlse.fr, chavanis@irsamc.ups-tlse.fr

Abstract. Using a generalization of the Madelung transformation, we derive the hydrodynamic representation of the Klein-Gordon-Einstein equations in the weak field limit. We consider a complex self-interacting scalar field with an arbitrary potential of the form $V(|\varphi|^2)$. We compare the results with simplified models in which the gravitational potential is introduced by hand in the Klein-Gordon equation, and assumed to satisfy a (generalized) Poisson equation. Nonrelativistic hydrodynamic equations based on the Schrödinger-Poisson equations or on the Gross-Pitaevskii-Poisson equations are recovered in the limit $c \to +\infty$.

1. Introduction

Scalar fields (SF) play an important role in particle physics, astrophysics, and cosmology [1]. Their evolution is usually described by the Klein-Gordon (KG) equation [2, 3] which can be viewed as a relativistic extension of the Schrödinger equation [4]. The KG equation describes spin-0 particles (bosons) that can be charged (complex SF) or neutral (real SF). The coupling between the KG equation and gravity through the Einstein equations, leading to the Klein-Gordon-Einstein (KGE) equations, was first considered in the context of boson stars [5, 6, 7]. It has also been proposed that dark matter (DM) halos may be made of a SF described by the KGE equations (see, e.g., [8, 9, 10] for recent reviews). Actually, at the galactic scale, the Newtonian limit is valid so that DM halos can be described by the Schrödinger-Poisson (SP) equations or by the Gross-Pitaevskii-Poisson (GPP) equations. In that case, the wave function $\psi(\vec{x}, t)$ describes a Bose-Einstein condensate (BEC) at $T = 0$, and the self-interaction of the bosons is measured by their scattering length a_s. Therefore, DM halos could be gigantic quantum objects made of BECs. The wave properties of bosonic DM may stabilize the system against gravitational collapse, providing halo cores and sharply suppressing small-scale linear power. This may solve the problems of the cold dark matter (CDM) model such as the cusp problem and the missing satellite problem. The scalar field dark matter (SFDM) model and the BEC dark matter (BECDM) model, also called ΨDM models, have received much attention in the last years.

Since DM may be a SF, it is of considerable interest to study the cosmological implications of this scenario. The cosmological evolution of a spatially homogeneous noninteracting real SF described by the KGE equations competing with baryonic matter, radiation and dark energy was considered by Matos *et al.* [11]. They found that real SFs display fast oscillations but that, on the mean, they reproduce the cosmological predictions of the standard ΛCDM model. The study of perturbations was considered by Suárez and Matos [12] for a self-interacting real SF described by the Klein-Gordon-Poisson (KGP) equations and by Magaña *et al.* [13] for a noninteracting real SF described by the KGE equations. These studies show that the perturbations can grow in the linear regime, leading, in the nonlinear regime, to the formation of structures corresponding to DM halos. This is in agreement with the early work of Khlopov *et al.* [14] who studied the Jeans instability of a relativistic SF in a static background. The case of a complex self-interacting SF representing BECDM was considered by Chavanis [15] in the context of Newtonian cosmology. His study is based on the GPP equations. Harko [16] and Chavanis [15] developed an approximate relativistic BEC cosmology and found that the perturbations grow faster in BECDM as compared to ΛCDM. Recently, Li *et al.* [17] developed an exact relativistic cosmology for a complex self-interacting SF/BEC based on the KGE equations. They studied the evolution of the homogeneous background and showed that the Universe undergoes three successive phases: a stiff matter era, followed by a radiation era due to the SF (that exists only for self-interacting SFs), and finally a matter era similar to CDM.

Instead of working directly in terms of a SF, we can adopt a fluid approach and work with hydrodynamic equations. In the case of the Schrödinger equation, this hydrodynamic approach was introduced by Madelung [18]. He showed that the Schrödinger equation is equivalent to the Euler equations for an irrotational fluid with an additional quantum potential arising from the finite value of $\hbar$. This hydrodynamic representation has been used by Böhmer and Harko [19] and by Chavanis [15, 20, 21] among others in the case of BECDM and in the case of BEC stars [22]. This hydrodynamic approach has been generalized by Suárez and Matos [12, 23] in the context of the KG equation. They used it to study the formation of structures in the Universe, assuming that DM is in the form of a fundamental SF with a $\lambda\varphi^4$ potential.

In the works [12, 23], the SF is taken to be real and the gravitational potential is introduced by hand in the KG equation, and assumed to be determined by the classical Poisson equation where the source is the rest-mass density ρ. This leads to the KGP equations. However, this treatment is not self-consistent since it combines relativistic and nonrelativistic equations. In these Proceedings, we derive the hydrodynamic representation of a complex SF coupled to gravity through the Einstein equations in the weak field approximation. In this way, we develop a self-consistent relativistic treatment. Throughout this work, we use the conformal Newtonian gauge which takes into account metric perturbations up to first order. We consider only scalar perturbations. This is sufficient if we are interested in calculating observational consequences of the SF dynamics in the linear regime. We compare our results with those obtained from the heuristic KGP equations. In these Proceedings, we develop the basic formalism and provide the main equations for an arbitrary self-interaction potential of the form $V(|\varphi|^2)$. Details and applications of our formalism can be found in our research papers [24, 25].

2. The conformal Newtonian gauge

We consider the KGE equations in the weak field limit $\Phi/c^2 \ll 1$. The equations that we derive are valid at the order $O(\Phi/c^2)$. We work with the conformal Newtonian gauge which is a perturbed form of the Friedmann-Lemaître-Robertson-Walker (FLRW) line element [26]. We consider the simplest form of the Newtonian gauge, only taking into account scalar perturbations which are the ones that contribute to the formation of structures in cosmology. We neglect anisotropic stresses. We assume that the Universe is flat in agreement with the observations of the cosmic microwave background (CMB). Under these conditions, our line element is given by

$$ds^2 = c^2\left(1+2\frac{\Phi}{c^2}\right)dt^2 - a(t)^2\left(1-2\frac{\Phi}{c^2}\right)\delta_{ij}dx^i dx^j, \tag{1}$$

where $\Phi/c^2 \ll 1$. In this metric, $\Phi(\vec{x},t)$ represents the gravitational potential of classical Newtonian gravity and $a(t)$ is the (dimensionless) scale factor.

3. The Lagrangian of the scalar field

We assume that DM can be described by a complex SF which is a continuous function of space and time defined at each point by $\varphi(x^\mu) = \varphi(x,y,z,t)$. The action of the relativistic SF is

$$S_\varphi = \int d^4x\sqrt{-g}\mathcal{L}_\varphi, \tag{2}$$

where $\mathcal{L}_\varphi = \mathcal{L}_\varphi(\varphi, \varphi^*, \partial_\mu\varphi, \partial_\mu\varphi^*)$ is the Lagrangian density and $g = \det(g_{\mu\nu})$ is the determinant of the metric tensor. We adopt the following generic Lagrangian density

$$\mathcal{L}_\varphi = \frac{1}{2}g^{\mu\nu}\partial_\mu\varphi^*\partial_\nu\varphi - \frac{m^2c^2}{2\hbar^2}|\varphi|^2 - V(|\varphi|^2) \tag{3}$$

which is written for a metric signature $(+,-,-,-)$. The second (quadratic) term is the rest-mass term. The potential $V(|\varphi|^2)$ takes into account the self-interaction of the SF. In certain applications, it is relevant to consider a quartic potential of the form

$$V(|\varphi|^2) = \frac{m^2}{2\hbar^4}\lambda|\varphi|^4, \tag{4}$$

where λ is the self-interaction coupling strength. If the SF describes a BEC at $T = 0$, the quartic potential (4) models the two-particle self-interaction.[1] In that case, m represents the mass of the bosons and λ is a constant related to the s-wave scattering length of the bosons a_s (measured in the center of mass frame) by $\lambda = 4\pi a_s\hbar^2/m$. This corresponds to the first Born approximation. The potential can then be written as

$$V(|\varphi|^2) = \frac{2\pi a_s m}{\hbar^2}|\varphi|^4. \tag{5}$$

4. The energy-momentum tensor

The energy-momentum tensor of the SF is

$$T_{\mu\nu} = 2\frac{\partial\mathcal{L}_\varphi}{\partial g^{\mu\nu}} - g_{\mu\nu}\mathcal{L}_\varphi. \tag{6}$$

For the generic Lagrangian (3), it takes the form

$$T_{\mu\nu} = \frac{1}{2}(\partial_\mu\varphi^*\partial_\nu\varphi + \partial_\nu\varphi^*\partial_\mu\varphi) - g_{\mu\nu}\left[\frac{1}{2}g^{\rho\sigma}\partial_\rho\varphi^*\partial_\sigma\varphi - \frac{m^2c^2}{2\hbar^2}|\varphi|^2 - V(|\varphi|^2)\right]. \tag{7}$$

By analogy with the energy-momentum tensor of a perfect fluid, the energy density and the pressure tensor of the SF are defined by $\epsilon = T_0^0$ and $P_i^j = -T_i^j$. The conservation of the energy-momentum tensor, which results from the Noether theorem, writes $D_\nu T^{\mu\nu} = 0$.

[1] It is a good approximation to ignore higher-order interactions when the boson gas is dilute, i.e., when the particle self-interaction range is much smaller than the mean interparticle distance.

5. The Klein-Gordon equation

The equation of motion for the SF can be obtained from the principle of least action. Imposing $\delta S_\varphi = 0$ for arbitrary variations $\delta\varphi$ and $\delta\varphi^*$, we obtain the Euler-Lagrange equation

$$D_\mu \left[\frac{\partial \mathcal{L}_\varphi}{\partial(\partial_\mu \varphi)^*} \right] - \frac{\partial \mathcal{L}_\varphi}{\partial \varphi^*} = 0, \tag{8}$$

where D is the covariant derivative. For the generic Lagrangian (3), this leads to the KG equation

$$\Box\varphi + \frac{m^2 c^2}{\hbar^2}\varphi + 2V(|\varphi|^2)_{,\varphi^*} = 0, \tag{9}$$

where $\Box$ is the d'Alembertian operator

$$\Box \equiv D_\mu (g^{\mu\nu}\partial_\nu) = \frac{1}{\sqrt{-g}}\partial_\mu(\sqrt{-g}\, g^{\mu\nu}\partial_\nu) \tag{10}$$

and

$$V(|\varphi|^2)_{,\varphi^*} = \frac{dV}{d|\varphi|^2}\varphi. \tag{11}$$

Computing the d'Alembertian (10) with the Newtonian gauge, we obtain the KG equation

$$\frac{1}{c^2}\frac{\partial^2\varphi}{\partial t^2} + \frac{3H}{c^2}\frac{\partial\varphi}{\partial t} - \frac{1}{a^2}\left(1+\frac{4\Phi}{c^2}\right)\Delta\varphi - \frac{4}{c^4}\frac{\partial\Phi}{\partial t}\frac{\partial\varphi}{\partial t} + \left(1+\frac{2\Phi}{c^2}\right)\frac{m^2c^2}{\hbar^2}\varphi + 2\left(1+2\frac{\Phi}{c^2}\right)V_{,\varphi^*} = 0, \tag{12}$$

where $H = \dot{a}/a$ is the Hubble constant. Using the expression (7) of the energy-momentum tensor, the energy density and the pressure are given by

$$\epsilon = T_0^0 = \frac{1}{2c^2}\left(1-\frac{2\Phi}{c^2}\right)\left|\frac{\partial\varphi}{\partial t}\right|^2 + \frac{1}{2a^2}\left(1+\frac{2\Phi}{c^2}\right)|\vec{\nabla}\varphi|^2 + \frac{m^2c^2}{2\hbar^2}|\varphi|^2 + V(|\varphi|^2), \tag{13}$$

$$P = -\frac{1}{3}(T_1^1 + T_2^2 + T_3^3) = \frac{1}{2c^2}\left(1-\frac{2\Phi}{c^2}\right)\left|\frac{\partial\varphi}{\partial t}\right|^2 - \frac{1}{6a^2}\left(1+\frac{2\Phi}{c^2}\right)|\vec{\nabla}\varphi|^2 - \frac{m^2c^2}{2\hbar^2}|\varphi|^2 - V(|\varphi|^2). \tag{14}$$

6. The Einstein equations

The Einstein equations write

$$R_{\mu\nu} - \frac{1}{2}g_{\mu\nu}R = \frac{8\pi G}{c^4}T_{\mu\nu}, \tag{15}$$

where $R_{\mu\nu}$ is the Ricci tensor, R is the Ricci scalar, and G is Newton's gravitational constant [27]. The conservation of the energy-momentum tensor is automatically included in the Einstein equations. The time-time component of the Einstein equations is

$$R_0^0 - \frac{1}{2}R = \frac{8\pi G}{c^4}T_0^0. \tag{16}$$

With the Newtonian gauge, it can be written as

$$\frac{\Delta\Phi}{4\pi G a^2} = \frac{\epsilon}{c^2} - \frac{3H^2}{8\pi G} + \frac{3H}{4\pi G c^2}\left(\frac{\partial\Phi}{\partial t} + H\Phi\right). \tag{17}$$

Using the expression (13) of the energy density ϵ, which represents the time-time component of the energy-momentum tensor, we get

$$\frac{\Delta\Phi}{4\pi G a^2}=\frac{1}{2c^4}\left(1-\frac{2\Phi}{c^2}\right)\left|\frac{\partial\varphi}{\partial t}\right|^2+\frac{1}{2a^2c^2}\left(1+\frac{2\Phi}{c^2}\right)|\vec{\nabla}\varphi|^2+\frac{m^2}{2\hbar^2}|\varphi|^2$$
$$+\frac{1}{c^2}V(|\varphi|^2)-\frac{3H^2}{8\pi G}+\frac{3H}{4\pi Gc^2}\left(\frac{\partial\Phi}{\partial t}+H\Phi\right). \quad (18)$$

Eqs. (12) and (18) form the KGE equations.

7. Spatially homogeneous scalar field

For a spatially homogeneous SF with $\varphi_b(\vec{x},t)=\varphi_b(t)$ and $\Phi_b(\vec{x},t)=0$, the KG equation (12) reduces to

$$\frac{1}{c^2}\frac{d^2\varphi_b}{dt^2}+\frac{3H}{c^2}\frac{d\varphi_b}{dt}+\frac{m^2c^2}{\hbar^2}\varphi_b+2V_{,\varphi_b^*}=0. \quad (19)$$

In that case, the energy-momentum tensor is diagonal and isotropic, $T^\mu_\nu=\mathrm{diag}(\epsilon_b,-P_b,-P_b,-P_b)$. The energy density $\epsilon_b(t)$ and the pressure $P_b(t)$ are given by

$$\epsilon_b=\frac{1}{2c^2}\left|\frac{d\varphi_b}{dt}\right|^2+\frac{m^2c^2}{2\hbar^2}|\varphi|^2+V(|\varphi_b|^2),\qquad P_b=\frac{1}{2c^2}\left|\frac{d\varphi_b}{dt}\right|^2-\frac{m^2c^2}{2\hbar^2}|\varphi|^2-V(|\varphi_b|^2). \quad (20)$$

From these equations, we obtain the continuity equation

$$\frac{d\epsilon_b}{dt}+3H(\epsilon_b+P_b)=0 \quad (21)$$

which is one of the Friedmann equations [27]. The other Friedmann equation is obtained from the Einstein equation (18) that reduces to

$$H^2=\frac{8\pi G}{3c^2}\epsilon_b. \quad (22)$$

This relation shows that the term $-3H^2/8\pi G=-\epsilon_b$ in the Einstein equation (17) plays the role of a neutralizing background.[2] From Eqs. (21) and (22), we easily obtain

$$\frac{\ddot{a}}{a}=-\frac{4\pi G}{3c^2}(\epsilon_b+3P_b). \quad (23)$$

8. The Gross-Pitaevskii-Einstein equations

The KG equation without self-interaction can be viewed as a relativistic generalization of the Schrödinger equation. Similarly, the KG equation with a self-interaction can be viewed as a relativistic generalization of the GP equation. In order to recover the Schrödinger and GP equations in the nonrelativistic limit $c\to+\infty$, we make the transformation [1]:

$$\varphi(\vec{x},t)=\frac{\hbar}{m}e^{-imc^2t/\hbar}\psi(\vec{x},t). \quad (24)$$

[2] In a static universe, the source of the gravitational potential is ϵ/c^2. In an expanding universe, the source of the gravitational potential is $(\epsilon-\epsilon_b)/c^2$, so that $\Phi=0$ when $\epsilon=\epsilon_b$. When we work in the comoving frame, the expansion of the universe amounts to subtracting a neutralizing background $-\epsilon_b/c^2$ to the density ϵ/c^2 [28], as in the Jellium model of plasma physics. This is the correct way to solve the Jeans swindle [29].

The prefactor $\hbar/m$ is justified in Appendix A. Mathematically, we can always make this change of variables. However, we emphasize that it is only in the nonrelativistic limit $c \to +\infty$ that ψ has the interpretation of a wave function, and that $|\psi|^2 = \rho$ has the interpretation of a rest-mass density. In the relativistic regime, ψ and $\rho = |\psi|^2$ do not have a clear physical interpretation. We will call them "pseudo wave function" and "pseudo rest-mass density". Nevertheless, it is perfectly legitimate to work with these variables and, as we shall see, the equations written in terms of these quantities take relatively simple forms that generalize naturally the nonrelativistic ones.

Substituting Eq. (24) into Eqs. (12) and (18), we obtain

$$i\hbar\frac{\partial\psi}{\partial t} - \frac{\hbar^2}{2mc^2}\frac{\partial^2\psi}{\partial t^2} - \frac{3}{2}H\frac{\hbar^2}{mc^2}\frac{\partial\psi}{\partial t} + \frac{\hbar^2}{2ma^2}\left(1+\frac{4\Phi}{c^2}\right)\Delta\psi - m\Phi\psi$$
$$-\left(1+\frac{2\Phi}{c^2}\right)m\frac{dV}{d|\psi|^2}\psi + \frac{3}{2}i\hbar H\psi + \frac{2\hbar^2}{mc^4}\frac{\partial\Phi}{\partial t}\left(\frac{\partial\psi}{\partial t} - \frac{imc^2}{\hbar}\psi\right) = 0, \tag{25}$$

$$\frac{\Delta\Phi}{4\pi Ga^2} = \left(1-\frac{\Phi}{c^2}\right)|\psi|^2 + \frac{\hbar^2}{2m^2c^4}\left(1-\frac{2\Phi}{c^2}\right)\left|\frac{\partial\psi}{\partial t}\right|^2 + \frac{\hbar^2}{2a^2m^2c^2}\left(1+\frac{2\Phi}{c^2}\right)|\vec{\nabla}\psi|^2$$
$$+\frac{1}{c^2}V(|\psi|^2) - \frac{\hbar}{mc^2}\left(1-\frac{2\Phi}{c^2}\right)\mathrm{Im}\left(\frac{\partial\psi}{\partial t}\psi^*\right) - \frac{3H^2}{8\pi G} + \frac{3H}{4\pi Gc^2}\left(\frac{\partial\Phi}{\partial t} + H\Phi\right). \tag{26}$$

Eq. (25) can be interpreted as a generalized Schrödinger equation (in the absence of self-interaction) or as a generalized GP equation (in the presence of self-interaction). It is coupled to the Einstein equation (26). The energy density and the pressure can be written as

$$\epsilon = \frac{\hbar^2}{2m^2c^2}\left(1-\frac{2\Phi}{c^2}\right)\left|\frac{\partial\psi}{\partial t}\right|^2 - \frac{\hbar}{m}\left(1-\frac{2\Phi}{c^2}\right)\mathrm{Im}\left(\frac{\partial\psi}{\partial t}\psi^*\right)$$
$$+\frac{\hbar^2}{2a^2m^2}\left(1+\frac{2\Phi}{c^2}\right)|\vec{\nabla}\psi|^2 + \frac{1}{2}c^2\left(1-\frac{2\Phi}{c^2}\right)|\psi|^2 + \frac{1}{2}c^2|\psi|^2 + V(|\psi|^2), \tag{27}$$

$$P = \frac{\hbar^2}{2m^2c^2}\left(1-\frac{2\Phi}{c^2}\right)\left|\frac{\partial\psi}{\partial t}\right|^2 - \frac{\hbar}{m}\left(1-\frac{2\Phi}{c^2}\right)\mathrm{Im}\left(\frac{\partial\psi}{\partial t}\psi^*\right)$$
$$-\frac{\hbar^2}{6a^2m^2}\left(1+\frac{2\Phi}{c^2}\right)|\vec{\nabla}\psi|^2 + \frac{1}{2}c^2\left(1-\frac{2\Phi}{c^2}\right)|\psi|^2 - \frac{1}{2}c^2|\psi|^2 - V(|\psi|^2). \tag{28}$$

Eqs. (25) and (26) form the GPE equations. In the nonrelativistic limit $c \to +\infty$, they reduce to the GPP equations [15]:

$$i\hbar\frac{\partial\psi}{\partial t} + \frac{3}{2}i\hbar H\psi = -\frac{\hbar^2}{2ma^2}\Delta\psi + m\Phi\psi + m\frac{dV}{d|\psi|^2}\psi, \tag{29}$$

$$\frac{\Delta\Phi}{4\pi Ga^2} = |\psi|^2 - \frac{3H^2}{8\pi G}. \tag{30}$$

For the quartic potential (5), we have

$$V(|\psi|^2) = \frac{2\pi a_s\hbar^2}{m^3}|\psi|^4. \tag{31}$$

9. The hydrodynamic representation

Important characteristics of the system are revealed by reformulating the KGE equations in the form of hydrodynamic equations. This can be done at the level of the GPE equations (25)-(26) via the Madelung transformation [18]. To that purpose, we write the pseudo wave function ψ as

$$\psi(\vec{x},t) = \sqrt{\rho(\vec{x},t)}e^{iS(\vec{x},t)/\hbar}, \tag{32}$$

where $\rho = |\psi|^2$ plays the role of a pseudo rest-mass density and S plays the role of a pseudo action. Following Madelung, we also define a pseudo velocity field as

$$\vec{v}(\vec{x},t) = \frac{\vec{\nabla} S}{ma}, \tag{33}$$

where the scale factor a has been introduced in order to take into account the expansion of the Universe. We note that this velocity field is irrotational.

Substituting Eqs. (32)-(33) into the GPE equations (25)-(26), and separating real and imaginary parts, we obtain the system of hydrodynamic equations

$$\frac{\partial\rho}{\partial t} + 3H\rho + \frac{1}{a}\vec{\nabla}\cdot(\rho\vec{v}) = \frac{1}{mc^2}\frac{\partial}{\partial t}\left(\rho\frac{\partial S}{\partial t}\right) + \frac{3H\rho}{mc^2}\frac{\partial S}{\partial t} + \frac{4\rho}{mc^4}\frac{\partial\Phi}{\partial t}\left(mc^2 - \frac{\partial S}{\partial t}\right) - \frac{4\Phi}{ac^2}\vec{\nabla}\cdot(\rho\vec{v}), \tag{34}$$

$$\begin{aligned}\frac{\partial S}{\partial t} + \frac{(\vec{\nabla}S)^2}{2ma^2} = &-\frac{\hbar^2}{2mc^2}\frac{\frac{\partial^2\sqrt{\rho}}{\partial t^2}}{\sqrt{\rho}} + \left(1+\frac{4\Phi}{c^2}\right)\frac{\hbar^2}{2ma^2}\frac{\Delta\sqrt{\rho}}{\sqrt{\rho}} - \frac{2\Phi}{mc^2a^2}(\vec{\nabla}S)^2\\ &-m\Phi - \left(1+\frac{2\Phi}{c^2}\right)mh(\rho) + \frac{1}{2mc^2}\left(\frac{\partial S}{\partial t}\right)^2 - \left(3H - \frac{4}{c^2}\frac{\partial\Phi}{\partial t}\right)\frac{\hbar^2}{4mc^2\rho}\frac{\partial\rho}{\partial t},\end{aligned} \tag{35}$$

$$\begin{aligned}\frac{\partial\vec{v}}{\partial t} + H\vec{v} + \frac{1}{a}(\vec{v}\cdot\vec{\nabla})\vec{v} = &-\frac{\hbar^2}{2am^2c^2}\vec{\nabla}\left(\frac{\frac{\partial^2\sqrt{\rho}}{\partial t^2}}{\sqrt{\rho}}\right) + \frac{\hbar^2}{2m^2a^3}\vec{\nabla}\left[\left(1+\frac{4\Phi}{c^2}\right)\frac{\Delta\sqrt{\rho}}{\sqrt{\rho}}\right]\\ &-\frac{1}{a}\vec{\nabla}\Phi - \frac{1}{\rho a}\vec{\nabla}p - \frac{2}{ac^2}\vec{\nabla}(h\Phi) - \frac{2}{ac^2}\vec{\nabla}(\Phi v^2) + \frac{1}{2am^2c^2}\vec{\nabla}\left[\left(\frac{\partial S}{\partial t}\right)^2\right]\\ &-\frac{3\hbar^2}{4am^2c^2}H\vec{\nabla}\left(\frac{1}{\rho}\frac{\partial\rho}{\partial t}\right) + \frac{\hbar^2}{am^2c^4}\vec{\nabla}\left(\frac{\partial\Phi}{\partial t}\frac{1}{\rho}\frac{\partial\rho}{\partial t}\right),\end{aligned} \tag{36}$$

$$\begin{aligned}\frac{\Delta\Phi}{4\pi Ga^2} = &\left(1-\frac{\Phi}{c^2}\right)\rho + \frac{\hbar^2}{2m^2c^4}\left(1-\frac{2\Phi}{c^2}\right)\left[\frac{1}{4\rho}\left(\frac{\partial\rho}{\partial t}\right)^2 + \frac{\rho}{\hbar^2}\left(\frac{\partial S}{\partial t}\right)^2\right]\\ &+\frac{\hbar^2}{2a^2m^2c^2}\left(1+\frac{2\Phi}{c^2}\right)\left[\frac{1}{4\rho}(\vec{\nabla}\rho)^2 + \frac{\rho}{\hbar^2}(\vec{\nabla}S)^2\right]\\ &+\frac{1}{c^2}V(\rho) - \frac{1}{mc^2}\left(1-\frac{2\Phi}{c^2}\right)\rho\frac{\partial S}{\partial t} - \frac{3H^2}{8\pi G} + \frac{3H}{4\pi Gc^2}\left(\frac{\partial\Phi}{\partial t} + H\Phi\right),\end{aligned} \tag{37}$$

where $h(\rho) = V'(\rho)$ is a pseudo enthalpy and $p(\rho)$ is a pseudo pressure defined by the relation $h'(\rho) = p'(\rho)/\rho$ [20]. It is explicitly given by $p(\rho) = \rho h(\rho) - \int h(\rho)\,d\rho$, i.e.,

$$p(\rho) = \rho V'(\rho) - V(\rho). \tag{38}$$

The pseudo velocity of sound is $c_s^2 = p'(\rho) = \rho V''(\rho)$. For the quartic potential (5), we have

$$V(\rho) = \frac{2\pi a_s \hbar^2}{m^3}\rho^2, \qquad h(\rho) = \frac{4\pi a_s \hbar^2}{m^3}\rho, \qquad p(\rho) = \frac{2\pi a_s \hbar^2}{m^3}\rho^2, \qquad c_s^2 = \frac{4\pi a_s \hbar^2}{m^3}\rho. \tag{39}$$

The pseudo pressure is given by a polytropic equation of state of index $\gamma = 2$ which is quadratic. We note that this equation of state coincides with the equation of state of a nonrelativistic self-interacting BEC [30]. This coincidence is not obvious because Eqs. (34)-(37) are valid in the relativistic regime. The interpretation of this equation of state is, however, not direct because ρ and p are a pseudo density and a pseudo pressure that coincide with the real density and the real pressure of a BEC only in the nonrelativistic limit $c \to +\infty$.

The energy density and the pressure can be written in terms of hydrodynamic variables as

$$\epsilon = \frac{\hbar^2}{2m^2c^2}\left(1-\frac{2\Phi}{c^2}\right)\left[\frac{1}{4\rho}\left(\frac{\partial\rho}{\partial t}\right)^2 + \frac{\rho}{\hbar^2}\left(\frac{\partial S}{\partial t}\right)^2\right] + \frac{\hbar^2}{2a^2m^2}\left(1+\frac{2\Phi}{c^2}\right)\left[\frac{1}{4\rho}(\vec{\nabla}\rho)^2 + \frac{\rho}{\hbar^2}(\vec{\nabla}S)^2\right]$$
$$-\left(1-\frac{2\Phi}{c^2}\right)\frac{\rho}{m}\frac{\partial S}{\partial t} + \frac{1}{2}\left(1-\frac{2\Phi}{c^2}\right)\rho c^2 + \frac{1}{2}\rho c^2 + V(\rho), \tag{40}$$

$$P = \frac{\hbar^2}{2m^2c^2}\left(1-\frac{2\Phi}{c^2}\right)\left[\frac{1}{4\rho}\left(\frac{\partial\rho}{\partial t}\right)^2 + \frac{\rho}{\hbar^2}\left(\frac{\partial S}{\partial t}\right)^2\right] - \frac{\hbar^2}{6a^2m^2}\left(1+\frac{2\Phi}{c^2}\right)\left[\frac{1}{4\rho}(\vec{\nabla}\rho)^2 + \frac{\rho}{\hbar^2}(\vec{\nabla}S)^2\right]$$
$$-\left(1-\frac{2\Phi}{c^2}\right)\frac{\rho}{m}\frac{\partial S}{\partial t} + \frac{1}{2}\left(1-\frac{2\Phi}{c^2}\right)\rho c^2 - \frac{1}{2}\rho c^2 - V(\rho). \tag{41}$$

We note that, in general, the pressure P defined by Eq. (41) is different from the pressure p defined by Eq. (38).[3] However, they coincide for a homogeneous SF in the regime where the SF oscillations are faster than the Hubble expansion [24, 25].

The hydrodynamic equations (34)-(37) have a clear physical interpretation. Eq. (34), corresponding to the imaginary part of the GPE equations, is the continuity equation. We note that $\int \rho\, d^3x$ is not conserved in the relativistic regime. However, Eq. (34) is consistent with the conservation of the charge of a SF (see section 10 and Appendix B). Eq. (35), corresponding to the real part of the GPE equations, is the Bernoulli or Hamilton-Jacobi equation. Eq. (36), obtained by taking the gradient of Eq. (35), is the momentum equation. Eq. (37) is the Einstein equation. We stress that the hydrodynamic equations (34)-(37) are equivalent to the GPE equations (25)-(26) which are themselves equivalent to the KGE equations (12) and (18). In the nonrelativistic limit $c \to +\infty$, we recover the quantum Euler-Poisson equations [15]:

$$\frac{\partial\rho}{\partial t} + 3H\rho + \frac{1}{a}\vec{\nabla}\cdot(\rho\vec{v}) = 0, \tag{42}$$

$$\frac{\partial S}{\partial t} + \frac{(\vec{\nabla}S)^2}{2ma^2} = \frac{\hbar^2}{2ma^2}\frac{\Delta\sqrt{\rho}}{\sqrt{\rho}} - m\Phi - mh(\rho), \tag{43}$$

$$\frac{\partial\vec{v}}{\partial t} + H\vec{v} + \frac{1}{a}(\vec{v}\cdot\vec{\nabla})\vec{v} = \frac{\hbar^2}{2m^2a^3}\vec{\nabla}\left(\frac{\Delta\sqrt{\rho}}{\sqrt{\rho}}\right) - \frac{1}{a}\vec{\nabla}\Phi - \frac{1}{\rho a}\vec{\nabla}p, \tag{44}$$

$$\frac{\Delta\Phi}{4\pi Ga^2} = \rho - \frac{3H^2}{8\pi G}. \tag{45}$$

These equations can also be written in terms of the density contrast $\delta = (\rho - \rho_b)/\rho_b$ [15].

[3] In our formalism, p represents the pressure arising from the self-interaction of the bosons (scattering) while P is the pressure of the SF defined by analogy with the pressure of an ideal fluid whose energy momentum tensor writes $T_{\mu\nu} = (P+\epsilon)u_\mu u_\nu/c^2 - Pg_{\mu\nu}$.

Journal of Physics: Conference Series **654** (2015) 012008 doi:10.1088/1742-6596/654/1/012008

10. Cosmological evolution of a spatially homogeneous scalar field

We consider the evolution of a universe induced solely by a spatially homogeneous SF. In the comoving frame, we have $\rho(\vec{x},t) = \rho_b(t)$, $\vec{v}_b(\vec{x},t) = \vec{0}$, $\Phi_b(\vec{x},t) = 0$, and $S_b(\vec{x},t) = S_b(t)$. We introduce the notation $E(t) = -dS_b/dt$ which can be considered as the time-dependent energy of the spatially homogeneous SF in the comoving frame. The pseudo wave function of the SF is $\psi_b(\vec{x},t) = \psi_b(t) = \sqrt{\rho_b(t)}e^{-(i/\hbar)\int E(t)\,dt}$. Using Eq. (24), we have $\varphi_b(\vec{x},t) = \varphi_b(t) = \frac{\hbar}{m}\sqrt{\rho_b(t)}e^{-(i/\hbar)[mc^2t+\int E(t)\,dt]}$ so the total energy of the SF, including its rest mass, is $E_{\rm tot}(t) = E(t) + mc^2$.

For a spatially homogeneous SF, the hydrodynamic equations (34)-(37) reduce to

$$\frac{d\rho_b}{dt} + 3H\rho_b = -\frac{1}{mc^2}\frac{d}{dt}(\rho_b E) - \frac{3H\rho_b}{mc^2}E, \tag{46}$$

$$\left(\frac{E}{2mc^2}+1\right)E = \frac{\hbar^2}{2mc^2}\frac{\frac{d^2\sqrt{\rho_b}}{dt^2}}{\sqrt{\rho_b}} + mh(\rho_b) + \frac{3H\hbar^2}{4mc^2\rho_b}\frac{d\rho_b}{dt}, \tag{47}$$

$$\frac{3H^2}{8\pi G} = \rho_b + \frac{\hbar^2}{2m^2c^4}\left[\frac{1}{4\rho_b}\left(\frac{d\rho_b}{dt}\right)^2 + \frac{\rho_b}{\hbar^2}E^2\right] + \frac{1}{c^2}V(\rho_b) + \frac{E}{mc^2}\rho_b. \tag{48}$$

In terms of the total energy $E_{\rm tot}(t) = E(t) + mc^2$, the equation of continuity (46) becomes

$$\frac{1}{\rho_b}\frac{d\rho_b}{dt} + \frac{3}{a}\frac{da}{dt} + \frac{1}{E_{\rm tot}}\frac{dE_{\rm tot}}{dt} = 0. \tag{49}$$

It can be rewritten as a conservation law:

$$\frac{d}{dt}(E_{\rm tot}\,\rho_b a^3) = 0. \tag{50}$$

Therefore, the total energy is exactly given by

$$\frac{E_{\rm tot}}{mc^2} = \frac{Qm}{\rho_b a^3}, \tag{51}$$

where Q is a constant. This conservation law was found by Gu and Hwang [31] directly from the KG equation. It can be shown that $Q = \int J^0\, d^3x$ represents the conserved charge density of the complex SF (see Appendix B). The energy density and the pressure of a homogeneous SF are

$$\epsilon_b = \frac{\hbar^2}{8m^2c^2}\frac{1}{\rho_b}\left(\frac{d\rho_b}{dt}\right)^2 + \frac{\rho_b}{m}E\left(1+\frac{E}{2mc^2}\right) + \rho_b c^2 + V(\rho_b), \tag{52}$$

$$P_b = \frac{\hbar^2}{8m^2c^2}\frac{1}{\rho_b}\left(\frac{d\rho_b}{dt}\right)^2 + \frac{\rho_b}{m}E\left(1+\frac{E}{2mc^2}\right) - V(\rho_b). \tag{53}$$

Equations (47), (48) and (51) determine the complete evolution of a universe induced by a spatially homogeneous SF. Working directly on the homogeneous KG equation with a quartic self-interaction potential, Li *et al.* [17] have shown that a universe filled with a relativistic complex SF first undergoes an intrinsic stiff matter era, followed by a radiation era due to its self-interaction, before finally entering in the matter era. The stiff matter era occurs when the

SF oscillations are slower than the Hubble expansion while the radiation and matter eras occur when the SF oscillations are faster than the Hubble expansion. These different regimes can be recovered from the hydrodynamic equations (46)-(48) [24].

In the nonrelativistic limit $c \to +\infty$, Eqs. (46)-(48) reduce to

$$\frac{d\rho_b}{dt} + 3H\rho_b = 0, \qquad E = mh(\rho_b), \qquad \frac{3H^2}{8\pi G} = \rho_b. \tag{54}$$

We find that $\rho_b \propto a^{-3}$, $a \propto t^{2/3}$ and $\rho_b = 1/(6\pi G t^2)$ (Einstein-de Sitter solution), so the homogeneous SF/BEC behaves as CDM. For the quartic potential (5), we have $E(t) = 4\pi a_s \hbar^2 \rho_b/m^2 = 2a_s\hbar^2/3Gm^2t^2$ and $S_b(t) = 2a_s\hbar^2/3Gm^2t + C$.

11. Generalized Klein-Gordon-Poisson equations

In this section, we consider a simplified model in which we introduce the gravitational potential $\Phi(\vec{x}, t)$ in the ordinary KG equation by hand, as an external potential, and assume that this potential is produced by the SF itself via a generalized Poisson equation in which the source is the energy density ϵ. This leads to the generalized KGP equations. We then show that these equations can be rigorously justified from the KGE equations in the limit $\Phi/c^2 \to 0$. However, this simplified model is not sufficient to study the evolution of the perturbations in the linear relativistic regime since it precisely neglects terms of order Φ/c^2.

We consider the FLRW metric that describes an isotropic and homogeneous expanding background. The line element in the comoving frame is

$$ds^2 = g_{\mu\nu}dx^\mu dx^\nu = c^2dt^2 - a(t)^2\delta_{ij}dx^i dx^j. \tag{55}$$

For this metric, the d'Alembertian operator (10) writes

$$\Box = \frac{1}{c^2}\frac{\partial^2}{\partial t^2} + \frac{3H}{c^2}\frac{\partial}{\partial t} - \frac{1}{a^2}\Delta. \tag{56}$$

In order to take the self-gravity of the SF into account, we introduce a Lagrangian of interaction that couples the gravitational potential $\Phi(\vec{x}, t)$ to the scalar field $\varphi(\vec{x}, t)$ according to

$$\mathcal{L}_{\rm int} = -\frac{m^2}{\hbar^2}\Phi|\varphi|^2. \tag{57}$$

The total Lagrangian of the system (SF + gravity) is given by $\mathcal{L} = \mathcal{L}_\varphi + \mathcal{L}_{\rm int}$. The equation of motion resulting from the stationarity of the total action $S = S_\varphi + S_{\rm int}$, obtained by writing $\delta S = 0$, is the KG equation

$$\Box\varphi + \frac{m^2c^2}{\hbar^2}\varphi + 2V(|\varphi|^2)_{,\varphi^*} + \frac{2m^2}{\hbar^2}\Phi\varphi = 0, \tag{58}$$

where the d'Alembertian operator is given by Eq. (56) and the gravitational potential $\Phi(\vec{x}, t)$ acts here as an external potential. The energy density and the pressure, defined from the diagonal part of the energy-momentum tensor (7), are given by

$$\epsilon = \frac{1}{2c^2}\left|\frac{\partial\varphi}{\partial t}\right|^2 + \frac{1}{2a^2}|\vec{\nabla}\varphi|^2 + \frac{m^2c^2}{2\hbar^2}|\varphi|^2 + V(|\varphi|^2), \tag{59}$$

$$P = \frac{1}{2c^2}\left|\frac{\partial\varphi}{\partial t}\right|^2 - \frac{1}{6a^2}|\vec{\nabla}\varphi|^2 - \frac{m^2c^2}{2\hbar^2}|\varphi|^2 - V(|\varphi|^2). \tag{60}$$

Eq. (58) is the ordinary KG equation for a SF in an external potential $\Phi(\vec{x},t)$ in an expanding background. We now state that $\Phi(\vec{x},t)$ is actually the gravitational potential produced by the SF itself. We phenomenologically assume that the gravitational potential is determined by a generalized Poisson equation of the form

$$\frac{\Delta\Phi}{4\pi G a^2}=\frac{1}{c^2}(\epsilon-\epsilon_b) \tag{61}$$

in which the source of the gravitational potential is the energy density ϵ of the SF (more precisely, its deviation from the homogeneous background density $\epsilon_b(t)$). Using Eq. (59) for the energy density of a SF, and recalling the Friedmann equation (22), the generalized Poisson equation can be written as

$$\frac{\Delta\Phi}{4\pi G a^2}=\frac{1}{2c^4}\left|\frac{\partial\varphi}{\partial t}\right|^2+\frac{1}{2a^2c^2}|\vec{\nabla}\varphi|^2+\frac{m^2}{2\hbar^2}|\varphi|^2+\frac{1}{c^2}V(|\varphi|^2)-\frac{3H^2}{8\pi G}. \tag{62}$$

Eqs. (58) and (62) form the generalized KGP equations. They have been introduced in an *ad hoc* manner but they can be rigorously justified from the KGE equations (12) and (18) in the limit $\Phi/c^2\to 0$ (which, of course, is different from the nonrelativistic limit $c\to+\infty$). We see that the gravitational potential Φ appears in the KG equation (58) due to the cancelation of c^2 in the product $\Phi/c^2\times c^2$ in Eq. (12). Therefore, we do not have to introduce Φ by hand: the generalized KGP equations can be obtained from the KGE equations by simply neglecting terms of order Φ/c^2 in these equations. Similarly, the equations related to the generalized KGP equations can be obtained from the ones related to the KGE equations by neglecting terms of order Φ/c^2. For example, the generalized GPP equations write

$$i\hbar\frac{\partial\psi}{\partial t}-\frac{\hbar^2}{2mc^2}\frac{\partial^2\psi}{\partial t^2}-\frac{3}{2}H\frac{\hbar^2}{mc^2}\frac{\partial\psi}{\partial t}+\frac{\hbar^2}{2ma^2}\Delta\psi-m\Phi\psi-m\frac{dV}{d|\psi|^2}\psi+\frac{3}{2}i\hbar H\psi=0, \tag{63}$$

$$\frac{\Delta\Phi}{4\pi G a^2}=|\psi|^2+\frac{\hbar^2}{2m^2c^4}\left|\frac{\partial\psi}{\partial t}\right|^2+\frac{\hbar^2}{2a^2m^2c^2}|\vec{\nabla}\psi|^2+\frac{1}{c^2}V(|\psi|^2)-\frac{\hbar}{mc^2}\mathrm{Im}\left(\frac{\partial\psi}{\partial t}\psi^*\right)-\frac{3H^2}{8\pi G}. \tag{64}$$

The energy density and the pressure are given by

$$\epsilon=\frac{\hbar^2}{2m^2c^2}\left|\frac{\partial\psi}{\partial t}\right|^2-\frac{\hbar}{m}\mathrm{Im}\left(\frac{\partial\psi}{\partial t}\psi^*\right)+\frac{\hbar^2}{2a^2m^2}|\vec{\nabla}\psi|^2+c^2|\psi|^2+V(|\psi|^2), \tag{65}$$

$$P=\frac{\hbar^2}{2m^2c^2}\left|\frac{\partial\psi}{\partial t}\right|^2-\frac{\hbar}{m}\mathrm{Im}\left(\frac{\partial\psi}{\partial t}\psi^*\right)-\frac{\hbar^2}{6a^2m^2}|\vec{\nabla}\psi|^2-V(|\psi|^2). \tag{66}$$

The corresponding hydrodynamic equations write

$$\frac{\partial\rho}{\partial t}+3H\rho+\frac{1}{a}\vec{\nabla}\cdot(\rho\vec{v})=\frac{1}{mc^2}\frac{\partial}{\partial t}\left(\rho\frac{\partial S}{\partial t}\right)+\frac{3H\rho}{mc^2}\frac{\partial S}{\partial t}, \tag{67}$$

$$\frac{\partial S}{\partial t}+\frac{(\vec{\nabla}S)^2}{2ma^2}=-\frac{\hbar^2}{2mc^2}\frac{\frac{\partial^2\sqrt{\rho}}{\partial t^2}}{\sqrt{\rho}}+\frac{\hbar^2}{2ma^2}\frac{\Delta\sqrt{\rho}}{\sqrt{\rho}}-m\Phi-mh(\rho)+\frac{1}{2mc^2}\left(\frac{\partial S}{\partial t}\right)^2-\frac{3H\hbar^2}{4mc^2\rho}\frac{\partial\rho}{\partial t}, \tag{68}$$

$$\frac{\partial\vec{v}}{\partial t}+H\vec{v}+\frac{1}{a}(\vec{v}\cdot\vec{\nabla})\vec{v}=-\frac{\hbar^2}{2am^2c^2}\vec{\nabla}\left(\frac{\frac{\partial^2\sqrt{\rho}}{\partial t^2}}{\sqrt{\rho}}\right)+\frac{\hbar^2}{2m^2a^3}\vec{\nabla}\left(\frac{\Delta\sqrt{\rho}}{\sqrt{\rho}}\right)-\frac{1}{a}\vec{\nabla}\Phi-\frac{1}{\rho a}\vec{\nabla}p$$
$$+\frac{1}{2am^2c^2}\vec{\nabla}\left[\left(\frac{\partial S}{\partial t}\right)^2\right]-\frac{3\hbar^2}{4am^2c^2}H\vec{\nabla}\left(\frac{1}{\rho}\frac{\partial\rho}{\partial t}\right), \tag{69}$$

$$\frac{\Delta\Phi}{4\pi G a^2} = \rho + \frac{\hbar^2}{2m^2c^4}\left[\frac{1}{4\rho}\left(\frac{\partial\rho}{\partial t}\right)^2 + \frac{\rho}{\hbar^2}\left(\frac{\partial S}{\partial t}\right)^2\right] + \frac{\hbar^2}{2a^2m^2c^2}\left[\frac{1}{4\rho}(\vec{\nabla}\rho)^2 + \frac{\rho}{\hbar^2}(\vec{\nabla}S)^2\right] + \frac{1}{c^2}V(\rho) - \frac{1}{mc^2}\rho\frac{\partial S}{\partial t} - \frac{3H^2}{8\pi G}. \quad (70)$$

The energy density and the pressure can be written in terms of hydrodynamic variables as

$$\epsilon = \frac{\hbar^2}{2m^2c^2}\left[\frac{1}{4\rho}\left(\frac{\partial\rho}{\partial t}\right)^2 + \frac{\rho}{\hbar^2}\left(\frac{\partial S}{\partial t}\right)^2\right] + \frac{\hbar^2}{2a^2m^2}\left[\frac{1}{4\rho}(\vec{\nabla}\rho)^2 + \frac{\rho}{\hbar^2}(\vec{\nabla}S)^2\right] - \frac{\rho}{m}\frac{\partial S}{\partial t} + \rho c^2 + V(\rho), \quad (71)$$

$$P = \frac{\hbar^2}{2m^2c^2}\left[\frac{1}{4\rho}\left(\frac{\partial\rho}{\partial t}\right)^2 + \frac{\rho}{\hbar^2}\left(\frac{\partial S}{\partial t}\right)^2\right] - \frac{\hbar^2}{6a^2m^2}\left[\frac{1}{4\rho}(\vec{\nabla}\rho)^2 + \frac{\rho}{\hbar^2}(\vec{\nabla}S)^2\right] - \frac{\rho}{m}\frac{\partial S}{\partial t} - V(\rho). \quad (72)$$

This model correctly describes the homogeneous background for which $\Phi = 0$ but it is not sufficient to describe the evolution of the perturbations in the linear regime because we must precisely take into account the terms of order Φ/c^2 in this regime (except, of course, in the nonrelativistic limit $c \to +\infty$). Therefore, the use of the KGE equations is mandatory to study the evolution of the perturbations in the relativistic regime.

Remark: We could also assume that the gravitational potential is determined by a Poisson equation of the form

$$\frac{\Delta\Phi}{4\pi G a^2} = \rho \quad (73)$$

in which the source of the gravitational potential is the pseudo rest-mass density $\rho = |\psi|^2$ of the SF. Eqs. (58) and (73) form the KGP equations. This approximation has been considered in [12]. However, there is an inconsistency in coupling the relativistic KG equation (58) to the classical Poisson equation (73).

12. Conclusion

We have developed a formalism based on a relativistic SF described by the KGE equations in the weak field limit. We have transformed these equations into equivalent hydrodynamic equations. These equations are arguably more tractable than the KGE equations themselves. In the nonrelativistic limit, they reduce to the hydrodynamic equations directly obtained from the GPP equations [15]. Therefore, our formalism clarifies the connection between the relativistic and nonrelativistic treatments. We note that, in the relativistic regime, the hydrodynamic variables ψ, ρ, S, $\vec{v}$, p,... that we have introduced do not have a direct physical interpretation. It is only in the nonrelativistic limit $c \to +\infty$ that they coincide with the wave function, rest-mass density, action, velocity, and pressure. However, these variables are perfectly well-defined mathematically from the SF φ in any regime, and they are totally legitimate. Furthermore, in terms of these variables, the relativistic hydrodynamic equations take a relatively simple form that provides a natural generalization of the nonrelativistic hydrodynamic equations.

The complete study of these relativistic hydrodynamic equations is of considerable interest but it is, of course, of great complexity. In our research papers [24, 25], we have started their study in simple cases. We have checked that the hydrodynamic equations of the SFDM model reproduce the evolution of the homogeneous background obtained previously by Li *et al.* [17] directly from the KGE equations: a stiff matter era, followed by a radiation era (for a self-interacting SF), and a matter era. We have also started to study the evolution of the perturbations in the linear regime in a static and in an expanding universe. We have shown analytically and numerically that perturbations whose wavelength is below the Jeans length oscillate in time while perturbations whose wavelength is above the Jeans length grow linearly with the scale factor as in the CDM model. The growth of perturbations in the SF model is substantially faster than in the CDM model. We have also shown that general relativity attenuates or even prevents the growth of perturbations at very large scales, close to the horizon (Hubble length).

For physically relevant wavelengths, the nonrelativistic limit of our formalism is sufficient to describe the evolution of the perturbations in the matter era. However, even if relativistic corrections are weak in the matter era, we may wonder whether their effect could be detected in an era of precision cosmology. In particular, it would be interesting to see if one can observe differences between the KGE equations considered in these Proceedings and the heuristic KGP equations studied in the past, in which gravity is introduced by hand in the KG equations (see section 11).

In future works, it will be important to consider the nonlinear regime where structure formation actually occurs. In general, this problem must be addressed numerically. The hydrodynamic equations derived in these Proceedings may be very helpful because they may be easier to solve than the KGE equations. Therefore, numerical simulations using fluid dynamics should be developed in the future. As a first step, relativistic effects could be neglected and the nonrelativistic equations of Ref. [15] could be used. These equations are similar to the hydrodynamic equations of CDM except that they include a quantum potential (Heisenberg) and a pressure term (scattering) that avoid singularities at small scales [15]. In this respect, it may be recalled that the SP equations were introduced early by Widrow and Kaiser [32] as a procedure of small-scale regularization (a sort of mathematical trick) to prevent singularities in collisionless simulations of *classical* particles. In their approach, $\hbar$ is not the Planck constant, but rather an ajustable parameter that controls the spatial resolution. Their procedure may find a physical justification if DM is made of self-gravitating BECs [33].

Our relativistic formalism may have applications for other self-gravitating systems described by SFs or BECs besides DM. We can mention, for example, the case of boson stars [5, 6, 7] and the case of microscopic quantum black holes made of BECs of gravitons stuck at a quantum critical point [34, 35, 10]. It has also been proposed [22] that, because of their superfluid core, neutron stars could be BEC stars. Indeed, the neutrons (fermions) could form Cooper pairs and behave as bosons of mass $2m_n$. This idea may solve certain issues regarding the maximum mass of neutron stars. Finally, we may mention analog models of gravity in which BECs described by the GP equation or by the KG equation are used to simulate results of classical and quantum field theory in curved spacetime [36].

Acknowledgments
A. S. acknowledges CONACyT for the postdoctoral grant received.

Appendix A. The value of A

The GP equation is obtained from the KG equation by means of the transformation

$$\varphi = A e^{-imc^2 t/\hbar}\psi. \tag{A.1}$$

The constant A can be computed as follows. Substituting Eq. (A.1) into Eq. (13), we find that the energy density of the SF is given by

$$\frac{\epsilon}{c^2} = \frac{T_0^0}{c^2} = \frac{1}{2}\left(1-\frac{2\Phi}{c^2}\right)\frac{m^2}{\hbar^2}A^2|\psi|^2 + \frac{m^2}{2\hbar^2}A^2|\psi|^2 + \frac{A^2}{2c^4}\left(1-\frac{2\Phi}{c^2}\right)\left|\frac{\partial\psi}{\partial t}\right|^2$$
$$+\frac{A^2}{2a^2c^2}\left(1+\frac{2\Phi}{c^2}\right)|\vec{\nabla}\psi|^2 + \frac{1}{c^2}V(|\psi|^2) - \frac{mA^2}{\hbar c^2}\left(1-\frac{2\Phi}{c^2}\right)\mathrm{Im}\left(\frac{\partial\psi}{\partial t}\psi^*\right). \tag{A.2}$$

Taking the nonrelativistic limit $c\to+\infty$ of this equation, we obtain

$$\frac{\epsilon}{c^2} \to \frac{m^2A^2}{\hbar^2}|\psi|^2 = \frac{m^2A^2}{\hbar^2}\rho, \tag{A.3}$$

where $\rho = |\psi|^2$ is the rest-mass density. Since $\epsilon \sim \rho c^2$ in the nonrelativistic limit $c \to +\infty$, we find

$$A = \frac{\hbar}{m}. \tag{A.4}$$

Appendix B. Some comments about the Klein-Gordon equation

In this Appendix, we recall the difficulties associated with the interpretation of the KG equation. We also clarify the relation between the hydrodynamic representation of the KG equation and the conservation of the charge.

The fundamental equation of nonrelativistic quantum mechanics is the Schrödinger [4] equation

$$i\hbar\frac{\partial\psi}{\partial t} = -\frac{\hbar^2}{2m}\Delta\psi. \tag{B.1}$$

If we write the wave function under the form $\psi(\vec{x},t) = \sqrt{\rho(\vec{x},t)}e^{iS(\vec{x},t)/\hbar}$ and define the density ρ and the current $\vec{J}$ by

$$\rho = |\psi|^2, \tag{B.2}$$

$$\vec{J} = \rho\frac{\vec{\nabla}S}{m} = \frac{\hbar}{2im}(\psi^*\vec{\nabla}\psi - \psi\vec{\nabla}\psi^*), \tag{B.3}$$

where we have used $S = (\hbar/2i)\ln(\psi/\psi^*)$, we obtain the continuity equation

$$\frac{\partial\rho}{\partial t} + \nabla\cdot\vec{J} = 0. \tag{B.4}$$

This equation shows that the integral of the density $\int \rho d^3x$ is conserved. Furthermore, by definition, the density is positive: $\rho(\vec{x},t) \geq 0$. Therefore, $\rho(\vec{x},t)$ can be interpreted as a probability density.

In the Madelung hydrodynamic representation of the Schrödinger equation, we define the density ρ and the velocity $\vec{v}$ by

$$\rho = |\psi|^2, \qquad \vec{v} = \frac{\vec{\nabla}S}{m}. \tag{B.5}$$

These variables satisfy the continuity equation

$$\frac{\partial\rho}{\partial t} + \nabla\cdot(\rho\vec{v}) = 0. \tag{B.6}$$

Since $\vec{J} = \rho\vec{v}$ according to Eqs. (B.3) and (B.5), we immediately see the equivalence between Eqs. (B.4) and (B.6).

We now consider the KG [2, 3] equation

$$\frac{1}{c^2}\frac{\partial^2\varphi}{\partial t^2} - \Delta\varphi + \frac{m^2c^2}{\hbar^2}\varphi = 0 \tag{B.7}$$

which was initially proposed as a relativistic extension of the Schrödinger equation.[4] If we write the SF under the form $\varphi(\vec{x},t) = \sqrt{R(\vec{x},t)}e^{i\sigma(\vec{x},t)/\hbar}$ and introduce the quadricurrent

[4] The KG equation was actually discovered by Schrödinger before he found the equation that now bears his name [1]. The KG equation was also obtained by Fock [37], de Donder and van den Dungen [38], and Kudar [39].

$J^\mu = -R\partial^\mu\sigma/m = -(\hbar/2im)(\varphi^*\partial^\mu\varphi - \varphi\partial^\mu\varphi^*)$, we obtain the continuity equation $\partial_\mu J^\mu = 0$. The quadricurrent can be written as $J^\mu = (J^0, \vec{J})$ with

$$J^0 = -R\frac{\frac{1}{c}\frac{\partial\sigma}{\partial t}}{m} = -\frac{\hbar}{2imc}\left(\varphi^*\frac{\partial\varphi}{\partial t} - \varphi\frac{\partial\varphi^*}{\partial t}\right), \tag{B.8}$$

$$\vec{J} = R\frac{\vec{\nabla}\sigma}{m} = \frac{\hbar}{2im}(\varphi^*\vec{\nabla}\varphi - \varphi\vec{\nabla}\varphi^*). \tag{B.9}$$

If we introduce the KG density $\rho_{\rm KG} = J^0/c$, we can write the continuity equation as

$$\frac{\partial\rho_{\rm KG}}{\partial t} + \nabla\cdot\vec{J} = 0. \tag{B.10}$$

This equation shows that the integral of the KG density $\int \rho_{\rm KG} d^3x$ is conserved. However, the KG density $\rho_{\rm KG}(\vec{x},t)$ is *not* definite positive so it cannot be interpreted as a density probability. Another difficulty with the KG equation is that it allows negative kinetic energies as solution. Indeed, decomposing Eq. (B.7) into plane waves, we obtain two solutions of the form $\varphi_\pm(\vec{x},t) = A_\pm e^{i(\vec{p}\cdot\vec{x}-E_\pm t)/\hbar}$ with $E_\pm = \pm\sqrt{p^2c^2+m^2c^4}$.

The original difficulties encountered with the KG equation had interesting historical developments. Dirac [40] proposed another relativistic extension of the Schrödinger equation. The Dirac equation describes spin-1/2 massive particles (fermions) such as electrons and quarks. In Dirac's theory, the probability density is positive but negative energies are allowed. Dirac solved the problem of negative energies via the "hole" theory [41]. This leads to the concept of antiparticles that are related to negative energy eigenstates. Antimatter was unsuspected before Dirac's work. The first antiparticle, the positron, was experimentally discovered by Anderson [42] in 1932. On the other hand, Pauli and Weisskopf [43] proposed a new interpretation of the KG equation. They interpreted the KG density $\rho_{\rm KG} = J^0/c$ as a charge density which can be of arbitrary sign. If we define the charge density and the charge current by $\rho_e = meJ^0/c\hbar^2$ and $\vec{J}_e = me\vec{J}/\hbar^2$, where e is an elementary charge, we can rewrite the continuity equation (B.10) as

$$\frac{\partial\rho_e}{\partial t} + \nabla\cdot\vec{J}_e = 0. \tag{B.11}$$

This equation expresses the conservation of the total charge $Q_e = \int \rho_e\, d^3x$. Although the KG equation is not a successful relativistic generalization of the Schrödinger equation (initially introduced to describe the energy spectrum of the electron), this equation was resurrected in the context of quantum field theory where it was shown to describe spin-0 particles (bosons) such as π-mesons, pions, or the Higgs boson. In this interpretation, since $\rho_e^{(\pm)} = \pm(e|E_\pm|/\hbar^2c^2)|\varphi_\pm|^2$, φ_+ specifies particles with charge $+e$ and energy $+E$ while φ_- specifies antiparticles with the same mass but with charge $-e$ and energy $-E$. For a real SF the charge is zero since $\rho_e = 0$ and $\vec{J}_e = \vec{0}$ according to Eqs. (B.8) and (B.9).

In the hydrodynamic representation of the KG equation, we write

$$\varphi(\vec{x},t) = \frac{\hbar}{m}\sqrt{\rho(\vec{x},t)}e^{i[S(\vec{x},t)-mc^2t]/\hbar} \tag{B.12}$$

and define the density ρ and the velocity $\vec{v}$ by

$$\rho = \frac{m^2}{\hbar^2}|\varphi|^2, \qquad \vec{v} = \frac{\vec{\nabla}S}{m}. \tag{B.13}$$

These variables satisfy the continuity equation

$$\frac{\partial\rho}{\partial t}+\nabla\cdot(\rho\vec{v})=\frac{1}{mc^2}\frac{\partial}{\partial t}\left(\rho\frac{\partial S}{\partial t}\right). \tag{B.14}$$

The density $\rho(\vec{x},t)$ is positive by definition but it is not conserved. Therefore, $\rho(\vec{x},t)$ cannot be interpreted as a density probability. However, the continuity equation (B.14) can be rewritten as

$$\frac{\partial}{\partial t}\left(\rho\frac{E_{\rm tot}}{mc^2}\right)+\nabla\cdot(\rho\vec{v})=0, \tag{B.15}$$

where we have defined $E(\vec{x},t)=-\partial S/\partial t$ and $E_{\rm tot}(\vec{x},t)=mc^2+E(\vec{x},t)$. This equation implies the conservation of the integral $\int\rho E_{\rm tot}\,d^3x$. To show that this integral corresponds to the total charge of the SF, we note that $R=(\hbar^2/m^2)\rho$ and $\sigma=S-mc^2t$. Therefore,

$$\rho_e=-\frac{e}{c^2\hbar^2}R\frac{\partial\sigma}{\partial t}=-e\frac{\rho}{m}\frac{\frac{\partial S}{\partial t}-mc^2}{mc^2}=e\frac{\rho}{m}\left(\frac{E}{mc^2}+1\right)=e\frac{\rho}{m}\frac{E_{\rm tot}}{mc^2}, \tag{B.16}$$

$$\vec{J}_e=\frac{e}{\hbar^2}R\vec{\nabla}\sigma=e\frac{\rho}{m}\frac{\vec{\nabla}S}{m}=e\frac{\rho}{m}\vec{v}. \tag{B.17}$$

These relations first establish the equivalence between Eqs. (B.11), (B.14), and (B.15). Furthermore, they show that the total charge of the SF can be written as

$$Q_e=e\int\frac{\rho}{m}\frac{E_{\rm tot}}{mc^2}\,d^3x \tag{B.18}$$

in agreement with Eq. (B.15). Eq. (B.18) is consistent with the expression (51) of the charge of a spatially homogeneous SF in an expanding universe.

References

[1] A. Zee, *Quantum Field Theory in a Nutshell* (Princeton University Press, 2003)
[2] O. Klein, Z. Phys. **37**, 895 (1926)
[3] W. Gordon, Z. Phys. **40**, 117 (1926)
[4] E. Schrödinger, Ann. Phys. (Berlin) **386**, 109 (1926)
[5] D.J. Kaup, Phys. Rev. **172**, 1331 (1968)
[6] R. Ruffini, S. Bonazzola, Phys. Rev. **187**, 1767 (1969)
[7] M. Colpi, S.L. Shapiro, I. Wasserman, Phys. Rev. Lett. **57**, 2485 (1986)
[8] A. Suárez, V.H. Robles, T. Matos, Astrophys. Space Sci. Proc. **38**, 107 (2014)
[9] T. Rindler-Daller, P.R. Shapiro, Astrophys. Space Sci. Proc. **38**, 163 (2014)
[10] P.H. Chavanis, Self-gravitating Bose-Einstein condensates, in *Quantum Aspects of Black Holes*, edited by X. Calmet (Springer, 2015)
[11] T. Matos, A. Vázquez-González, J. Magaña, Mon. Not. R. Astron. Soc. **393**, 1359 (2009)
[12] A. Suárez, T. Matos, Mon. Not. R. Astron. Soc. **416**, 87 (2011)
[13] J. Magaña, T. Matos, A. Suárez, F. J. Sánchez-Salcedo, JCAP **10**, 003 (2012)
[14] M.Yu. Khlopov, B.A. Malomed, Ya.B. Zeldovich, Mon. Not. R. astr. Soc. **215**, 575 (1985)
[15] P.H. Chavanis, Astron. Astrophys. **537**, A127 (2012)
[16] T. Harko, Mon. Not. R. Astron. Soc. **413**, 3095 (2011)
[17] B. Li, T. Rindler-Daller, P.R. Shapiro, Phys. Rev. D **89**, 083536 (2014)
[18] E. Madelung, Zeit. F. Phys. **40**, 322 (1927)
[19] C.G. Böhmer, T. Harko, J. Cosmol. Astropart. Phys. **06**, 025 (2007)
[20] P.H. Chavanis, Phys. Rev. D **84**, 043531 (2011)
[21] P.H. Chavanis, L. Delfini, Phys. Rev. D **84**, 043532 (2011)
[22] P.H. Chavanis, T. Harko, Phys. Rev. D **86**, 064011 (2012)

[23] A. Suárez, T. Matos, Class. Quantum Grav. **31**, 045015 (2014)
[24] A. Suárez, P.H. Chavanis, Phys. Rev. D **92**, 023510 (2015)
[25] A. Suárez, P.H. Chavanis, in preparation
[26] C.-P. Ma, E. Bertschinger, Astrophys. J. **455**, 7 (1995)
[27] S. Weinberg, *Gravitation and Cosmology* (John Wiley, 1972)
[28] P.J.E. Peebles, *The Large-Scale Structure of the Universe* (Princeton University Press, 1980)
[29] J. Binney, S. Tremaine, *Galactic Dynamics* (Princeton University Press, 2008)
[30] F. Dalfovo, S. Giorgini, L.P. Pitaevskii, S. Stringari, Rev. Mod. Phys. **71**, 463 (1999)
[31] J.-A. Gu, W.-Y.P. Hwang, Phys. Lett. B **517**, 1 (2001)
[32] L.M. Widrow, N. Kaiser, Astrophys. J. Lett. **416**, L71 (1993)
[33] P.H. Chavanis, Phys. Rev. D **84**, 063518 (2011)
[34] G. Dvali, C. Gomez, Fortschr. Phys. **61**, 742 (2013)
[35] R. Casadio, A. Orlandi, J. High Energy Phys. **8**, 25 (2013)
[36] C. Barcelo, S. Liberati, M. Visser, Phys. Rev. A **68**, 053613 (2003)
[37] V. Fock, Z. Phys. **38**, 242 (1926)
[38] T. de Donder, H. van den Dungen, C. rend. Acad. Sci. (Paris) **183**, 22 (1926)
[39] J. Kudar, Ann. Phys. **81**, 632 (1926)
[40] P.A.M. Dirac, Proc. Royal Soc. A **117**, 610 (1928)
[41] P.A.M. Dirac, Proc. Royal Soc. A **126**, 360 (1930)
[42] C. Anderson, Phys. Rev. **43**, 491 (1933)
[43] W. Pauli, V. Weisskopf, Helv. Phys. Acta **7**, 709 (1934)